California Landscape

Winter in the California mountains gives an indication of the climate of thousands of years ago when glaciers extended down from the heights of the high Sierra Nevada. During the Great Ice Age, the last major episode in geologic time, the ice and the flooding of lands below the mountains changed the face of California extensively. Since then, agents of wind and water have worked to modify the landscape, carving it into spectacular beauty.

California Natural History Guide: 48

California Landscape
ORIGIN AND EVOLUTION

Mary Hill

Photos by Susan Moyer
Maps by Adrienne Morgan

UNIVERSITY OF CALIFORNIA PRESS
Berkeley   Los Angeles   London

California Natural History Guides
Arthur C. Smith, General Editor

Advisory Editorial Committee:
Mary Lee Jefferds
A. Starker Leopold
Robert Ornduff
Robert C. Stebbins

University of California Press
Berkeley and Los Angeles, California

University of California Press, Ltd.
London, England

Library of Congress Cataloging in Publication Data

Hill, Mary, 1923–
  California landscape.

  (California natural history guide;48)
  Bibliography: p.
  Includes index.
  1. Geomorphology—California.
I. Title.   II. Series.
GB428.C3H54   1984
551.4′09794   82-20256
ISBN 0-520-04831-8
ISBN 0-520-04849-0 (pbk.)

Printed in the United States of America

1 2 3 4 5 6 7 8 9

Contents

Sources and Acknowledgments

Figure 1 was redrawn from the pamphlet "Geologic Time," published by the U.S. Geological Survey. Figures 2, 4, and 18 are reprinted from *Geology of the Sierra Nevada*, by Mary Hill, University of California Press, 1975. Figure 3 is by courtesy of the U.S. Geological Survey. Figure 6 is from U.S. Geological Survey, folio 5, 1895. Figure 7 was taken from "Mt. Shasta, a Cascade Volcano," published in the *Journal of Geology*, vol. 40, 1932. Figures 8, 9, 10 and 12 were drawn by Elinor H. Rhodes, and are published by courtesy of the California Division of Mines and Geology. Figure 11 was also drawn by Mrs. Rhodes, after a photograph by B. F. Loomis. Figures 14, 15, and 16 were originally drawn by W. C. Putnam, first published in *The Geographical Review*, 1938. Figure 19 is a drawing by Patricia Edwards; figure 20 is another by Elinor H. Rhodes, from *Diving and Digging for Gold*, by Mary Hill, Naturegraph Publishers, 1974.

Figure 27 is reprinted from U.S. Geological Survey Professional Paper 941-A, 1975. Figure 31 is by W. M. Davis, first published in the *Scottish Geographical Magazine*, 1906. Figures 30 and 33 are by Alex Eng, from *Geology of the Sierra Nevada*. Figure 32 is from *Geologic History of the Sierra Nevada*, by François Matthes, published as U.S. Geological Survey Professional Paper 100, 1930. Figure 34 is from "Mt. Shasta, a Typical Volcano," by J. S. Diller, in *The Physiography of the United States*, National Geographic Society Monographs, vol. 1, 1896.

Figure 35 is from U.S. Geological Survey Annual Report for 1978; figure 39 was drawn by Ed Foster, courtesy of California Division of Mines and Geology. Figure 36 is by courtesy of the U.S. Geological Survey. Figure 42 was redrawn from an article by R. E. Wallace in *Proceedings of a Conference on Geologic Problems of the San Andreas Fault System*, Stanford University Publications in the Geological Sciences, vol. 11, 1968. Figure 41 is from U.S. Geological Survey Professional Paper 941-A. Figure 45 was redrawn from photographs in *American Practical Navigator*, originally by Nathaniel Bowditch, issued as U.S. Navy Hydrographic Office Publication No. 9. Figures 53 and 55 are by Tau Rho Alpha, courtesy of the U.S. Geological Survey. Both of these are orthographic projections. Figure 68 is from *Geomorphology in Deserts* by Ronald U. Cooke and Andrew Warren, University of California Press, 1973. Figure 70 was redrawn and the quotation in its caption taken from *The Earth and Human Affairs*, by the Committee on Geological Sciences, Division of Earth Science, National Research Council–National Academy of Sciences, published by Canfield Press, 1972. Figure 71 was redrawn from *The Warm Desert Environment*, by Andrew Goudie and John Wilkinson, Cambridge University Press, 1977. All other figures were drawn by Hidekatsu Takada.

Maps 2 and 5 were redrawn from maps published by the California Division of Mines and Geology; map 3 was derived from maps prepared by the California Department of Parks and Recreation; map 8 is from data supplied by William Raub, C. Suzanne Brown, and Austin Post; map 9 was redrawn from the World Seismicity Map published by the U.S. Geological Survey and compiled by Arthur C. Tarr, 1974. Map 10 is from data compiled by the California Division of Mines and Geology. Map 11 was derived from information from the California Division of Mines and Geology and the U.S. Geological Survey. Map 13 was modified from maps published by the National Science Foundation in *Mosaic*, vol. 8, no. 1, January–February, 1977, based on Peveril Meigs's classification; map 14 was derived from information in the *National Atlas*, published by the U.S. Geological Survey; map 15 was modified from "Pleistocene Lakes in the Great Basin," by C. T. Snyder, George Hardman, and F. F. Zdnek, U.S. Geological Survey Miscellaneous Geologic Investigations Map I-416; map 16 was modified from "Pleistocene Lakes of Southeastern California," by Robert P. Blanc and George B. Cleveland, in *Mineral Information Service*, vol. 14, no. 5, May 1961, p. 5; maps 17 and 18 were compiled from U.S. Geological Survey topographic maps.

Table 1 was modified from the table on page 13 of *Earth*, by Frank Press and Raymond Siever, W. H. Freeman Co., San Francisco, 1974. Table 3 was derived from table 1 in "Resource appraisal of the Mt. Shasta Wilderness Study Area, Siskiyou County, California," by Robert L. Christiansen, Frank J. Kleinhampl, Richard J. Blakely, Ernest T. Tuchek, Fredrick L. Johnson, and Martin D. Conyac, published as Open-file report 77-250, U. S. Geological Survey, 1977. Table 4 was modified from "Volcanism in California" by Charles W. Chesterman, in *California Geology*, v. 24, no. 8, p. 141, August, 1971.

All photographs are by Susan Moyer except the one black-and-white photograph of Mount Lassen in eruption, which was taken by R. I. Myers.

The quotation on page 59 is by T. A. Jagger from Geological Society of America *Memoir* 21, 1947.

The quotation on page 197–98 is from *Roughing It* by Mark Twain, New York: Grosset and Dunlap, originally published in 1871, p. 132.

The quotation on pages 227–29 is from "Ecology of a Discovered Land," by David W. Mayfield, printed in *Pacific Discovery*, September–October, 1980, pages 12–20. It is published here by permission of the California Academy of Sciences.

I give particular thanks to Susan Moyer, Elisabeth Egenhoff, and the General Editor of this series, Arthur C. Smith, all of whom criticized the manuscript and encouraged me.

"The shape of things to come . . ."

California is large and varied. Within its borders are the highest peak and the lowest spot, the driest desert and the wettest mountain in the conterminous 48 states. Its long coastline has rugged cliffs and picturesque rocks, dangerous shoals and smooth, sandy beaches.

Volcanoes and Earth forces—including earthquakes—have built California's mountains. Volcanoes have added hot, molten rock from the Earth's interior, making high places where none were before. The Earth forces that produce earthquakes have lifted mountains, using the rocky material of which the land was made. Rocks that may have been formed in the deep sea are, through the movement of Earth, now risen to the heights. Even now, the mountains are growing by earthquake and volcano.

Other forces are changing the landscape, carving it into the spectacular shapes we cherish as the glorious scenery of California. Chief among these is water, in myriad forms. Rushing stream and plunging fall; roaring river and quiet slough; silent lake and pounding surf—all these as well as the raindrops themselves have sculpted California's land. Most striking artist of all was water working as great ice tongues that covered the mountains a few thousand years ago.

The chapters in this book are arranged so as to emphasize these processes of building and changing. The processes have acted through thousands and millions of years to create California as we know it, and are still changing it, even as we watch.

1 · THE MANY FACES OF TIME

The aim of earth science is to discover all there is to know about Earth, past and present. To do this, earth scientists borrow the techniques of chemistry and physics in order to decipher Earth's ingredients and their condition; they borrow from biology insight into its past inhabitants and some bits of their stories; they take from history the uses of chronology; and from engineering a measure of Earth's strengths and weaknesses. But they have a tool of their own: time. Time is the hallmark of geology. All the enlightenment geologists have derived with the aid of other sciences is used within a framework of time to unlock Earth's past and present, and, in some measure, to foretell its future.

Geology lacks one essence of the scientific method: repeatability. Because geology is inextricably involved with time, and because we have not yet learned to repeat time, we cannot reproduce long geological events to verify our conclusions. Time is the subject of the geological inquiry; time is the key; and time marks the limits—all of which leads us into difficult philosophical waters.

"Time, gentlemen!" is a curfew—a signal that one more day's business is complete. To the poet, time is a thief; to a musician, time is an ingredient. "I know I have to beat time when I learn music," said Alice. "Ah, that accounts for it," said the mad Hatter, "he won't stand beating. Now, if you'd only kept on good terms with him, he'd do almost anything for you."

1

In the Eastern view of the world, time is a hoop: the "ever-circling years" repeat themselves endlessly. In the Western world, we are accustomed to thinking of time as linear, "progressing" from one point to another. Nothing can repeat itself in this view, because even if all other parts were the same, time would have changed. Such a view makes it difficult for us to comprehend eternity, but it allows us to measure time as if it were units on a ruler.

The span of our lives is divided into units of time. We hear of, but can scarcely comprehend, very small units of time: chemical and physical reactions in which the lifetime of an element is less than the twinkling of an eye. With some difficulty, we can imagine being an insect, with a life expectancy of a few hours. It is easier for us to understand the precious seconds that television so carefully monitors; the hours of an office worker's day, the fading and blooming of garden flowers, or the seasonal march of a farm. The years pass in our own lives too quickly, and we grow older without noticing. All of these measures we are accustomed to. It is harder for us to look beyond our own lives: to differentiate Genghis Khan from Hamlet, for example, as historical figures before our own time often seem more fictional than real.

We are traveling in time as we study the stars from our vantage point here on Earth. Even though light travels at the unimaginable speed of 186,000 miles each second, space is so vast that starlight takes years to reach Earth. We see the star Alpha Centauri not as it is today, but as it was four years ago, and the Andromeda galaxy as it was during our own Ice Age two million years ago. "Our days," said an astronomer, "reside in the midst of a billion yesterdays."

The life span of a star, measured in billions of years, is as difficult for us to envision as the millisecond existence of an atomic particle. Our Earth is a youthful inhabitant of the heavens, one whose "days are as grass" in comparison to the span of the universe.

Even so, Earth's life span of about 4.5 billion years

counts many sunrises and sunsets, less than 1 percent of which the human race has been privileged to see. It is these 4.5 billion years—the lifetime of Earth—with which geology is concerned. The purpose of geology is to discover what happened during that time, how it happened, and why it happened. If from this we can predict what will happen, perhaps we can learn to live in harmony with nature, rather than destroying ourselves in an arrogant attempt to "conquer" her.

Although people have been making geologic observations throughout recorded history, it has only been within the past two centuries that we have deciphered the order of events in the life of Earth. As long ago as the fifth century B.C., the Greek historian Herodotus correctly observed that the Mediterranean Sea had once been much farther inland. He had no way of knowing that what he had observed was part of a very long story that involves the total rearranging of the geography of the planet Earth.

Thanks to thousands of workers these past 200 years, we now have at least a sketchy idea of Earth's story. The tools for deciphering this tale have progressed from the purely intellectual to the technological. We have not abandoned our early intellectual tools—we've merely added to them.

One of these tools is the idea that, by observing processes at work today we can deduce what has happened before. "The present is the key to the past" is how this proposition is generally stated. It may not be absolutely true; there may be something new under the sun, or there may have been forces or factors at work in the past that no longer operate. But unless we can show otherwise, it is a good starting point.

A second principle involves the observation of rock layers. It states that unless other forces have disrupted the Earth where layers of rock are stacked one upon the other, the youngest is on top. If one were making a bed, the last blanket to be laid on the stack would be on top; it would be the "youngest."

With these two principles in mind, it became possible to build a "time scale" by using a third principle derived from them—the idea of "correlation."

How this works is well demonstrated in the Four Corners area of the United States, where Colorado, Utah, New Mexico, and Arizona join in a point. This is canyonland country, where one can get lost vertically as well as horizontally. Deep in the Grand Canyon of the Colorado, a world-famous sequence of rocks has been gashed open by the Colorado River, providing inspiration for poets and artists and instruction for geologists. Within that layer cake of rocks, the two beds on top (and therefore, by our second postulate, the youngest) are a tan limestone that one stands on at the canyon rim and red cliffs one can look up to.

To the north in Zion National Park, Utah, it is possible to recognize the same rock layers again, although there is not the long sequence exposed beneath them here that there is in the Grand Canyon. Here are other layers above, including one bed of cross-bedded, light-colored rock (called the Navajo Sandstone); geologists who have studied it believe it to be ancient sand dunes turned to stone. This same cross-bedded sandstone can, in turn, be traced eastward to Canyonlands National Park, where more of the layers below are exposed, as well as several others above.

Within the layers above is one group of green and purple shale beds (the Morrison Formation), from which numerous dinosaur bones have been recovered elsewhere in the West. One of the many other places where the colorful Morrison beds are exposed is Mesa Verde National Park, in southern Colorado. There they provide a dash of brighter color in an otherwise tan landscape—a landscape dominated by sandstone beds on top of the Morrison. In niches in these overlying sandstone beds, the people who lived there as long ago as the thirteenth century built what we now call "condominiums" accessible only by narrow paths or chipped handholds in the nearly vertical cliffs.

Lowermost of these cliff-making sandstone beds is the

Dakota Sandstone (elsewhere important as a carrier of groundwater), which is traceable throughout much of the West. This bed is visible in the layers of rock at weirdly eroded Bryce Canyon National Park, where one can also trace beds downward to the cross-bedded Navajo sand dunes and upward to still younger beds.

In this way, it is possible to see how rocks are related to one another (correlated), and how a sequence can be built up from the oldest bed to the youngest, even though all beds may not be present in one place, and even though the builder may not have any idea of the actual age of any bed. This is how geologists have constructed Earth's history: bit by bit, taking a fact from here that can be related to one from there, and so on. Often the beds contain distinctive fossils that add further clarification to the proper ordering of Earth's story.

Gradually, scientists have built up a relative time scale, based on the relationships of beds to one another and on the fossils in them. The scale has some unfamiliar and difficult words in it; ones that were derived from Greek and Latin (common scientific languages) or from tribal and place names of Europe, together with a few American additions (Mississippian, for example).

This time scale has been extremely useful for scientists. It is still used today, although the actual age in years of many rocks is now known. It is possible to determine the actual ages of rocks (and thus of time periods) by several means, although more than one method can rarely be used on the same rock. Because they vary in their precision, most of the techniques provide dates within certain limits of accuracy rather than actual birthdays.

One group of methods depends upon radioactivity in rocks. A large handful of radioactive elements can be used for this purpose, including carbon, uranium, thorium, rubidium, potassium, radon, and hydrogen. The idea is this: radioactive elements (isotopes) "decay" by losing particles from their nuclei to form other, "daughter" elements (also

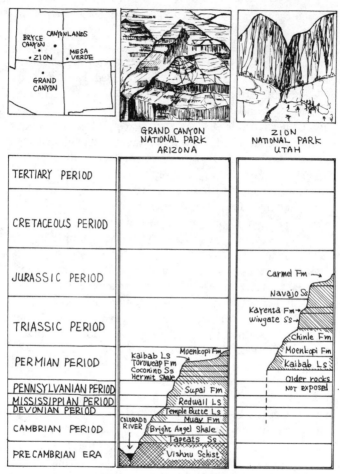

	GRAND CANYON NATIONAL PARK ARIZONA	ZION NATIONAL PARK UTAH
TERTIARY PERIOD		
CRETACEOUS PERIOD		
JURASSIC PERIOD		Carmel Fm → Navajo Ss
TRIASSIC PERIOD		Kayenta Fm → Wingate Ss → Chinle Fm
PERMIAN PERIOD	Kaibab Ls ⎯ Moenkopi Fm Toroweap Fm Coconino Ss Hermit Shale	Moenkopi Fm Kaibab Ls
PENNSYLVANIAN PERIOD	Supai Fm	Older rocks NOT EXPOSED
MISSISSIPPIAN PERIOD	Redwall Ls	
DEVONIAN PERIOD	Temple Butte Ls	
CAMBRIAN PERIOD	COLORADO RIVER ↓ Muav Fm Bright Angel Shale Tapeats Ss	
PRECAMBRIAN ERA	Vishnu Schist	

FIG. 1. How a sequence of rocks is built up, using information from widely separated areas. By matching the same rock beds from one area to another, the total stack of rocks can be deduced. From evi-

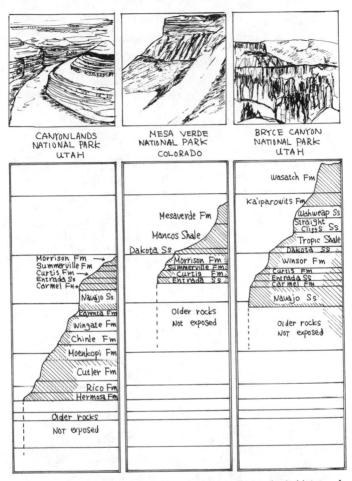

CANYONLANDS
NATIONAL PARK
UTAH

MESA VERDE
NATIONAL PARK
COLORADO

BRYCE CANYON
NATIONAL PARK
UTAH

Wasatch Fm

Ka'iparowits Fm

Wahweap Ss
Straight
Cliffs Ss

Tropic Shale

Dakota Ss

Winsor Fm

Curtis Fm
Entrada Ss
Carmel Fm

Navajo Ss

Older rocks
NOT exposed

Mesaverde Fm

Mancos Shale

Dakota Ss

Morrison Fm
Summerville Fm
Curtis Fm
Entrada Ss

Older rocks
Not exposed

Morrison Fm →
Summerville Fm
Curtis Fm →
Entrada Ss
Carmel Fm

Navajo Ss

Kayenta Fm

Wingate Fm

Chinle Fm

Moenkopi Fm

Cutler Fm

Rico Fm

Hermosa Fm

Older rocks
Not exposed

dence in that stack, scientists can determine the geologic history of the region.

RELATIVE GEOLOGIC TIME			ATOMIC TIME (in millions of years)
Era	Period		Epoch	
Cenozoic	Quaternary		Holocene	
			Pleistocene	
				— 2–3 —
	Tertiary		Pliocene	— 12 —
			Miocene	— 26 —
			Oligocene	— 37–38 —
			Eocene	— 53–54 —
			Paleocene	— 65 —
Mesozoic	Cretaceous		Late Early	— 136 —
	Jurassic		Late Middle Early	190–195
	Triassic		Late Middle Early	— 225 —
Paleozoic	Permian		Late Early	— 280 —
	Carbon-iferous Systems	Pennsyl-vanian	Late Middle Early	
		Mississip-pian	Late Early	— 345 —
	Devonian		Late Middle Early	— 395 —
	Silurian		Late Middle Early	430–440
	Ordovician		Late Middle Early	— 500 —
	Cambrian		Late Middle Early	— 570 —
Precambrian				— 3,600+ —

FIG. 2.

TABLE 1. Radioactive Elements Used to Determine the Ages of Rocks

Parent isotope	Daughter isotope	Half-life in years
Uranium-238	Lead-206	4.5 billion
Uranium-235	Lead-207	713 million
Thorium-232	Lead-208	14.1 billion
Rubidium-87	Strontium-87	50.0 billion
Potassium-40	Argon-40	1.3 billion
Carbon-14	Nitrogen-14	5.73 thousand

called isotopes). The rate of decay is spoken of as the "half life" of the isotope; that is, the time it takes half of the original isotope to become the new, daughter isotope. Most radioactive isotopes have very short half lives—seconds, days, or years. Some, however, decay more slowly. These are the elements useful in a geologic "atomic clock," for they can provide estimates of longer time past—closer in length to the eons of earth history.

Although the laboratory procedure for determining carbon-14 dates is quite complex, as are all radioactive methods, carbon has been widely used for dating historical, geological, and archaeological events. The reason it is so useful is that it can give the length of time since once-living things were alive. This is a curious contradiction: we normally think of the age of a living creature as its life span; but in using the atomic clock, scientists are interested in the life span of carbon after the organism itself has died.

Plants and animals, when alive, contain carbon 14; when they die, the carbon begins to disintegrate at the half-life rate of $5,730 \pm 40$ years. At this speed, the clock is wound so that it provides dates back to about 50,000 years ago. Because of laboratory uncertainties, dates beyond 30,000 years ago are not altogether reliable.

Another method of telling time past is to count tree rings. A casual glance at a tree stump shows that the cross section looks like a target, with rings circling around a center. Each ring represents the growth of the tree in a year, with narrow rings for drought years, and wider rings for years of plenty. If one knows when a tree was cut down, one has merely to count backward to the tree's youth. Each ring differs from its fellows slightly, just as each season had different weather. By examining rings from many different trees, scientists have been able to identify specific rings in different trees, and by using the principle of correlation, have been able to build up a long tree-ring chronology. In this way, they have been able to date events that occurred in the American Southwest, in Ancient Egypt, and during the days of the great glaciers, as well as to give us maps of past climates.

One very useful tree for this purpose is the Bristlecone Pine, a doughty inhabitant of wind-whipped mountain slopes. It is the second longest-lived tree on Earth; only the Mediterranean Cypress can grow to be older. The oldest Bristlecone Pine measured to date is 4,600 years old, exceeding the aged redwoods by many centuries, although it is a century younger than the oldest cypress.

Obviously, rings from such trees tell us long, long tales. And even more: recently scientists have used individual rings of the Bristlecone Pine to obtain carbon 14 dates. In this way, they know the precise year (by counting tree rings) and can verify the laboratory results of carbon dating.

Another use of living plants is in the calculation of dates by lichenometry—that is, the rate of growth of certain species of lichens. It is a difficult technique, because lichens, like people, grow at different rates during their lifetimes. However, in areas where there are no trees, such as near ice sheets, lichens are the next best alternative. To use the method, one has first to determine the regional growth rate of the plants involved (lichens are composite organisms consisting of algae and fungi), then apply it to specific cases. Obviously, this technique cannot be used to determine very ancient ages, as the lichens being measured are still alive.

The yearly habits of lakes furnish still another chronicle for use in research. "Varves" are pairs of light and dark layers laid down annually in lakes. If the varves are truly annual, and have not been disturbed by storms or earth movements, one would then have only to count by twos to arrive at correct dates in local history. Scandinavian scientists have set up a chronology using varves that date back 12,000 years.

Cave environments have given scientists an opportunity to use a technique called "thermoluminescence," which involves heating of samples in order to "leak" alpha particles from radioactive impurities. The alpha particles appear as light, which can be measured. The method is only useful in areas where the temperature has been constant, because the particles leak faster as the temperature is raised. If the temperature is constant (as in caves), then the leakage rate is also constant, and the amount remaining to be leaked gives a measure of the age.

Finally, a method of absolute time measurement that has recently been used extensively by geologists is one well known in other fields: study of the historical record. One can delve into old files for statements describing "the way it was," search out old drawings, or attempt to compile visual records to produce an intermittent or time-lapse effect. Such records can provide dates from the Stone Age onward, and are particularly useful in studying modern rates of change.

Even though we have learned to calculate time past, and have discovered the shape of some past events, we still need to learn to use time present. We do not need to beat him, as Alice confessed to, but we need to understand time present if we are to comprehend fully time past and time future.

We need to "handle" time if we are to estimate the rate at which events are happening. One way to handle time is to use it continuously, making a constant record of events as they occur. This is the way seismographs operate, and it is the way life operates.

Another way of using time is to quantify it—to use it in

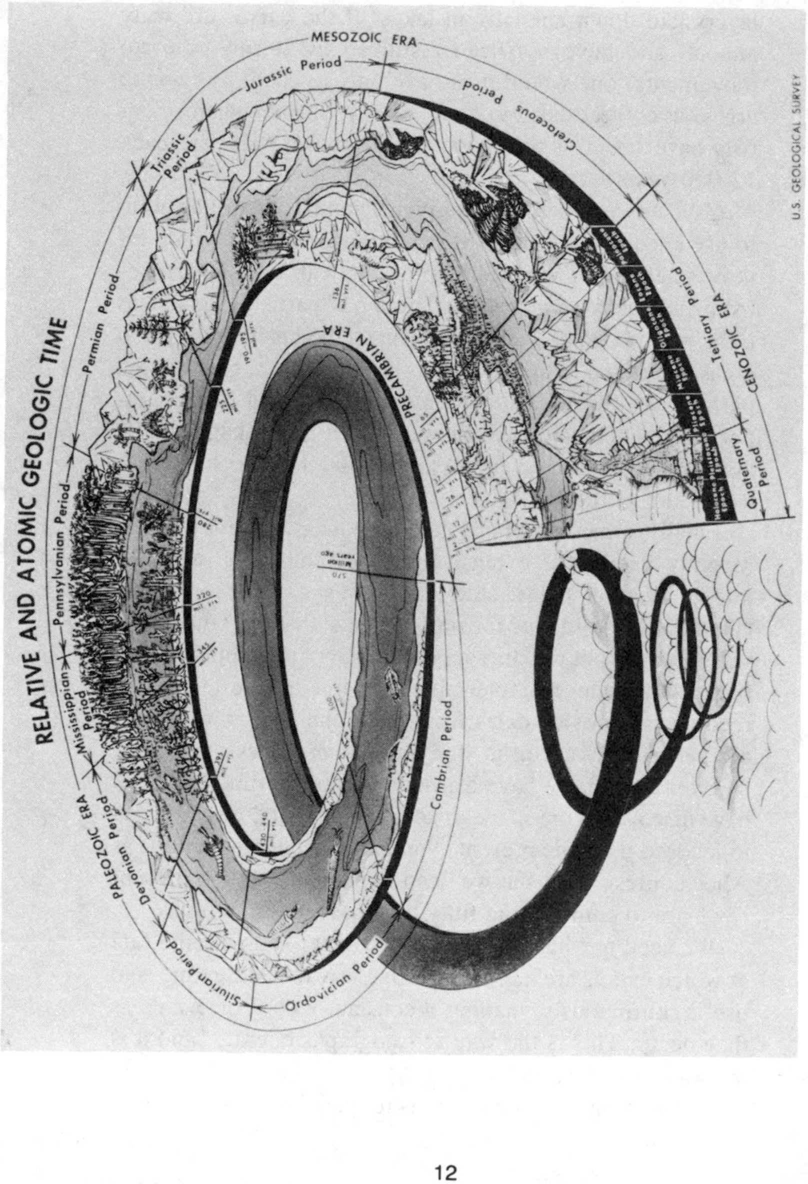
RELATIVE AND ATOMIC GEOLOGIC TIME
MESOZOIC ERA
Jurassic Period
Cretaceous Period
Triassic Period
Permian Period
Pennsylvanian Period
Mississippian Period
PALEOZOIC ERA
Devonian Period
Silurian Period
Ordovician Period
Cambrian Period
PRECAMBRIAN ERA
Tertiary Period
CENOZOIC ERA
Quaternary Period
U.S. GEOLOGICAL SURVEY

FIG. 3.

units or increments. This is the way stream gauges are read—once a month, or once a week. It is the way motion pictures are made—twenty-four frames each second, each frame an individual scene differing slightly from frames on either side. When the film is played back, the action seems continuous, just as when the stream gauge records are plotted, the line joining the points is continuous.

Or we can use time in discrete units, in which the interval of time is more significant than the relationship between the readings. Such a manipulation of time gives us acts in a play or actions of nature measured by the week or by the month or by the million years.

Or we can take "grab samples" of time, having no particular relationship to what has gone before or what will follow. Such sampling is like snipping one frame from each of several reels of film.

All these ways of dealing with time are efforts to answer fundamental questions about geologic events: what, when, how, and how often. These give us a picture of evolution through time—of how things change and how fast they change. It is becoming increasingly apparent that we need to know these things; that geology is not the study of yesterday solely for yesterday's sake, but for today's and tomorrow's as well.

2 · ROCKS, AND ROCKS FROM ROCKS

Earth, beneath a thin veneer of water, soil, plants, and human construction, is made up of rock. For that reason, the study of rocks is the foundation of geology.

If one remembers that some rocks are stacked in layers, heaped upon one another like layers of a cake, and that the uppermost one is the youngest (the last one to be stacked), one might suppose that wherever we see bedded rocks, they should be horizontal. This is not so, as a trip down nearly any California highway will reveal. Layers are at all angles; some flat, some upright, some twisted, some bent. If we assume that all the rocks were once horizontal, then we are led to the conclusion that some force has moved them. In other words, Earth has changed. It is this idea of change that is central to geology: through time, Earth changes. How it changes and why it changes is the scientist's task to decipher.

I have said, "If we assume that the rocks were once horizontal," but what makes us think that these layered rocks were once horizontal? If, indeed, the present is the key to the past, then where on today's Earth can we find a modern counterpart—horizontal layers of incipient rock?

Observation of lakes, rivers, and seas shows us that layers of sand, gravel, mud, and shells accumulate on their bottoms; it does not take much imagination to suppose that these are the conditions under which bedded rocks form. Here, then, scientists arrived at the first great group of rocks: bedded rocks, such as sandstone, shale, and limestone.

If one were to shake a glass of muddy water, gradually the mud—sediment—would fall to the bottom as a layer, just as it does in Earth's water bodies. From this sediment came the name for this group of rocks: sedimentary (see fig. 4). It takes its origin from the Latin *sedere*, meaning to sit; it is given to these rocks because they are "seated" on the surface of the Earth, accumulated there in beds that later are turned to stone.

Accumulated. Accumulated from what? Again, observation tells us that they can be accumulated from such sources as pebbles of other rocks, grains of sand, bodies of animals and plants, and soil, to say nothing of such modern additions as broken bottles, cans, and other trash. How all this came to be available as sediment is another story.

Among the pebbles of other rocks that one might identify in sedimentary rocks are pieces of lava, which, we now know, solidifies from molten rock poured from volcanoes or volcanic fissures. Lava cools quickly, in a geologic sense, to form hard rock that is very finely grained—basalt, andesite, or rhyolite. Each grain is actually a microscopic crystal. Sometimes lava is quenched so quickly that no crystals form at all. Volcanic glass is the result. At other times, volcanoes may "boil over" as candy on a kitchen stove may. The cooled rock froth is called pumice.

Volcanic rocks, like bedded sedimentary rocks, are ones that we can see in the process of formation, and therefore be reasonably certain that our idea of their origin is correct. There are other rocks we have never seen forming—at least, to our knowledge—so that our understanding of them comes from experiment and inference. One group of such rocks is chemically identical with lava but, unlike lava, it is made up of easily visible mineral crystals. Geologists reason that these coarser grained rocks also were once molten, but that they cooled far enough underground, and therefore slowly enough, for larger crystals to form. Because they

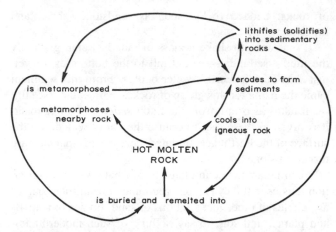

FIG. 4. The rock cycle. The three classes of rock—igneous, sedimentary, and metamorphic—grade into one another, and are, through time, transformed one into another. Hot fluid material, for example, may cool and harden into igneous rock, either on the surface of the earth as lava or beneath it, perhaps as granite. The lava flow or the granite may be worn to fragments by wind and weather to accumulate as sedimentary layers, which in turn lithify into rock. The sedimentary rock may itself be eroded into fragments (reworked) to become new layers of sediment and again lithify into sedimentary rock. Perhaps it is then buried deeply and metamorphosed or remelted. Metamorphic rock may, in turn, be eroded to form sediments or remelted to start the cycle over.

cool underground, they are called "plutonic" rocks, for Pluto, god of the underworld.

Volcanic and plutonic rocks (also called "extrusive" and "intrusive" rocks) are commonly grouped together under one heading, "igneous," derived from the same root as the word "ignite." The great bulk of Earth is igneous rock.

Any of these rocks, sedimentary or igneous, might be seen as pebbles in accumulating sedimentary layers. Most of the layers, as they accumulate, are horizontal or nearly so; yet most layers that we see in California are not horizontal. Something has changed them—twisted and turned them, squeezed them, stretched them.

Not only do rocks lithify, or harden, as the layers of sediment must, if they are to become sedimentary rock, but some are also changed in form—"metamorphosed." Limestone may be altered by heat or pressure or both to marble, peat to coal, or shale to slate. For some rocks, that change is nearly as spectacular as the transformation of a caterpillar to a butterfly, while for others the change is so subtle as to be detectable only under a microscope.

By accumulation, solidification, and change, we have the three great groups of rocks: sedimentary, igneous, and metamorphic. The grouping reflects their origin, yet it is not often possible to tell the origin of a rock by merely looking at it. These ideas of rock origin have developed slowly in the past 200 years, after much evidence discovered by scientists the world over has been weighed.

Two centuries ago, scientists were debating the origin of basalt (a type of solidified lava), some contending that it was water laid. Today, a visit to an erupting Hawaiian volcano will quickly demonstrate the origin of basalt.

Although we now all agree on the origin of basalt, the origin of many other rocks is far from settled among scientists. Indeed, the names that rocks are called is not always agreed upon. For the purposes of this book, however, we will use names that we can arrive at by observation, rather than by laboratory examination or previous experience. Bear in mind, however, that the same rock may be known by many different names, all of which may be correct.

In dividing rocks into sedimentary, igneous, and metamorphic (based on their genesis), we are reaching conclusions that go beyond what we can actually see in the rock. This is the aim of geology: to interpret the story of the rock—and therefore the Earth—from observation.

First, however, we must make observations. It has taken a long time in the history of geology to learn what to look for in order to be able to make intelligent interpretations; as science progresses, and as there is need for more detail, different observations may be required, and it becomes necessary to refine previous work.

What shall we look for in rocks to distinguish them from one another? One clue has already been suggested: layering. Layering very likely indicates that the beds are sedimentary in nature. This is not always so; lava flows may be layered, and some metamorphic rocks are banded in such a way as to look layered. But other clues can quickly tell us whether or not the rock is part of a lava flow or is metamorphic.

Many rocks of the sedimentary rock class are composed of rounded grains held together by cement. They are given names according to the size of their grains: conglomerate (composed of cemented gravel), sandstone (from sand), and shale (from mud). Other rocks are chemically or organically derived; among them are limestone, salt, and chert.

To give names to igneous and metamorphic rocks requires a different set of observations. To separate rocks in these groups, one must be able to recognize a few individual minerals. Unlike rocks, minerals have specific attributes. All are chemical compounds that occur in nature, and although some of them grade into one another, most have definite names and definite physical and chemical properties.

The search for and identification of minerals is an engrossing and rewarding occupation and hobby. However, one need not be a professional mineralogist to identify rocks, as knowledge of a handful of minerals is sufficient to name most rocks. Indeed, about a dozen mineral species make up the bulk of the crust of the earth. Quartz, feldspar, mica, pyroxene, amphibole, calcite, garnet, olivine—these and a few others are enough to enable one to do a respectable job of field classification, although, of course, the more one knows of a subject, the more enjoyable it becomes.

Rocks do not change because we "identify" them, nor does our understanding of them deepen. We struggle with names so that we can talk with one another about rocks, and know that we are talking about the same things. It is true that in the process of identification our powers of observa-

TABLE 2. Minerals in the Earth's Crust

Mineral family	Percentage
Feldspar	58
Pyroxene, amphibole	13
Quartz	11
Mica, clay	10
Carbonate, oxide, sulfide, halide	3
Olivine	3
Other	2

tion are sharpened, but we must not lose sight of the fact that it is the interpretation of them, not the names, that we are seeking.

Why are sedimentary rocks made up, at least in part, of pieces of rocks? Where do the pieces come from?

The story of the accumulation of these pieces, large and small, is another chapter in the book of Earth changes. For once a rock has cooled from a molten mass or consolidated from an accumulation into hard rock, it may have become toughened, but it is far from everlasting. Earth's rocks, particularly those near the surface of the ground, are subject to the action of weathering and to the forces of erosion, both of which tend to break up rocks and wear down mountains.

Using the tools of water, ice, and wind, weathering works on the individual mineral grains, altering them chemically. In addition, water, because it is so easily transmutable under the physical conditions at the Earth's surface, breaks rocks apart as it changes from water to ice and back again. One has only to watch ice shatter bottles to realize how powerful a tool ice can be in the thin cracks within rocks.

One of the most common chemical reactions is the change from feldspar to clay, with the help of water. This is the reaction:

$$2KAlSi_3O_8 \quad + H_2O \quad + 2H \quad \rightarrow Al_2Si_3O_5(OH)_4$$

feldspar   + water   + hydrogen $\rightarrow$ clay

As most plutonic igneous rocks (which are the cores of many of our mountain ranges) contain a great deal of feldspar, this reaction, although slow compared with most chemical reactions, nevertheless is geologically quick to break apart the rocks of the mountains.

Almost as fast as the rock is broken apart, the forces of erosion carry it away. The newly broken particles move downhill by gravity, helped along by rains. Once they reach an area drained by streams, they can move more swiftly. Perhaps they are halted in a lake, where they can become part of a layer of lake sediment (such as the varves mentioned in the last chapter), or perhaps they make their way to the sea to become a layer of marine sediment. It may be that some grains are loosened to be picked up by the wind and piled into dunes. Or the clay and rock fragments may settle into a valley to become soil. Anything can happen, once the fragments are broken from the parent rock.

In general, the chemically resistant, inert grains, such as quartz, retain their identity for thousands of years—perhaps through several "lifetimes" as a sedimentary rock. These are the grains we commonly see in dunes and in hard sandstone. Other, softer or more chemically changeable minerals have long ago taken other forms.

It is this process of weathering and erosion, working on different kinds of rock, each having a different geologic history, operating at different rates in the varying climates of yesterday and today, which has created California's landforms. And these landforms are the face of California as we know it.

"Variety's the very spice of life."

3 · THE LAY OF THE LAND

California, like Shakespeare's man, plays many parts. One can see it as a piece of the universe, though how large a piece it is not possible to say with accuracy; or, one can see it as a part of the surface of the globe, of which it is about 2.5 percent of the total land area. Its existence, of course, is purely arbitrary, as no lines set it off in nature.

California is 415,195 square kilometers (158,693 square miles) in size, or about 40 million hectares (100 million acres). About half is in public ownership and half is privately owned. Of the total, 17 million hectares (43 million acres) is forest land; 15 million hectares (36 million acres) is used for agriculture; 8 million hectares (20 million acres) is desert and open space; and 1 million hectares (2 million acres) is densely urban.

The state extends from 32° to 40° North latitude, and from 114° to 124° West longitude. It is 1,325 kilometers (825 miles) in longest dimension (northwest to southeast), and 555 kilometers (345 miles) wide, measured at its widest point near Point Arguello. It has about 1,350 kilometers (840 miles) of coastline.

California's geographical center is 56 kilometers (35 miles) northeast of Madera. Its highest point is Mount Whitney in the Sierra Nevada, 4,418 meters (14,495 feet) in elevation; its lowest is Badwater, Death Valley National Monument, 86 meters (282 feet) below sea level. These two points are within sight of each other.

MAP 1. California counties and county seats.

Most of the state has two seasons, wet and dry. The wet season (mainly during the winter months) furnishes runoff to support many rivers. Average precipitation ranges from 3,000 mm (110 inches) in the northwest to less than 50 mm (2 inches) in Death Valley.

We can think of California as consisting chiefly of mountains and valleys, intertwined with rivers and dotted with lakes; or we can see it as cities grading into countryside, knitted together by highways.

Geographers have divided it into eight regions, dependent largely upon similar topography and similar climates. Landscape architects have divided it into nine provinces, each a unit because of a similar type of scenery. Geologists and geomorphologists (physiographers) cut the state into nine or ten units, depending upon whether similar geology or the similarity of landforms is the major consideration.

The backbone of the state is the great Sierra Nevada Range. If one uses the boundaries usually given for it—the Tehachapi Range on the south to the Cascade Range on the north—it is the longest continuous mountain bastion in the United States south of Alaska. Both the Appalachians in the east and the Rocky Mountains in the midcontinent are mountain systems, comprising several different ranges with different geologic histories. As the geology of the Klamath Mountains to the northwest is similar to the Sierra Nevada, they, too, can be considered part of the range, separated from the main mass by a covering of lava from the Cascades. Rocks of the Sierra Nevada also have affinities with granitic rocks in Idaho to the northeast, and with rocks in the Peninsular Ranges and in Mexico to the south.

Because the Sierra Nevada is a huge tilted block of rock much like a half-opened trap door in shape, the two sides do not have equal slopes. On the western side, the slope of the surface is generally not steep, averaging only about 2°, while on the east, it is precipitous.

Weather in the Sierra Nevada is exciting. In a wet year, the range receives an enormous amount of precipitation. Indeed, it is the great bulk of the Sierra, rising nearly three miles high, that intercepts the moisture from the Pacific, causing snow to fall, and preventing storms from continuing eastward. In this way, the Sierra acts as a snow catcher, its "shadow" causing the land to the east to be desert.

And snow it does. The unofficial North American snow record was set in the Sierra at Tamarack, in Alpine County

MAP 2. Geomorphic provinces of California.

in 1906, where, in that year, 23 meters (almost 76 feet!) fell. Tamarack also holds the record for the greatest snowfall in one calendar month—9,910 mm (390 inches—32.5 feet), set in January 1911, and for the greatest depth of snow on the ground at any one time (not counting snowdrifts)—1,145 cm (451 inches), nearly 37½ feet!

Spring thaws bring rushing rivers coursing down the

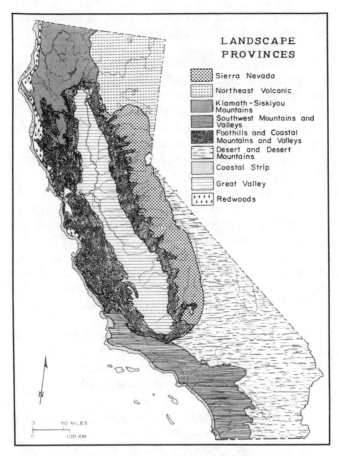

LANDSCAPE PROVINCES

Sierra Nevada

Northeast Volcanic

Klamath-Siskiyou Mountains

Southwest Mountains and Valleys

Foothills and Coastal Mountains and Valleys

Desert and Desert Mountains

Coastal Strip

Great Valley

Redwoods

MAP 3. Landscape provinces of California.

gentler western side of the Sierra Nevada. The rivers, flowing for thousands of years, have now cut more or less parallel east-west canyons through the western slope, making it much more rugged in a north-south direction than in an east-west direction.

Rocks of the Sierra Nevada range in age from early Paleozoic (roughly 500 million years old) to those being formed today in such places as mountain meadows, rivers, and lakes. The rock record holds the story of the Sierra in past ages, although it is not always easy to read.

Some parts of the Sierra Nevada were first uplifted from beneath the sea about 130 million years ago. Raising of the range has not yet ceased, however, as there have been spurts of uplift that have continued to the present day. While the mountains were being uplifted, the granite that forms the core of the range, together with solutions carrying precious and other metals, found their place—no doubt while still in a molten or fluid state.

During the long periods of erosion that followed uplift, streams and rivers, many following courses different from those of today, carried fragments of gold torn from the mountains. In places these streams were buried by volcanic ash and lava from Sierran volcanoes, entombing the riverbed, gold and all. The most vigorous part of the volcanic episode lasted 20 million years, and added 8,000 cubic kilometers (2,000 cubic miles) of rock to the landscape, while erosion continued to wear it down.

Just a few thousand years ago, the majestic scenery for which the Sierra is noted was formed by the slow scouring action of great glaciers. River valleys were transformed by the ice sheets, which, a short time ago, geologically, had covered the range with fingers of ice.

The Klamath Mountains, although geologically continuous with the Sierra Nevada, have had a different recent history. Because there were few glaciers in them, they were not carved into such a spectacular landscape. One of the reasons for the lack of glaciers is that the Klamaths were not uplifted as high as the Sierra. If one looks at a map, it is evident that the southern part of the Sierra proper is higher than the northern part, and that the Klamaths are, in general, still lower. A few places in the Klamaths were high enough to support glaciers. In addition, in the southern Sierra Nevada, the granite, which cooled deep underground, was lifted higher, and thereby exposed to erosion. As one progresses northward, less and less granite is exposed to view, because erosion has not cut down as deeply. Since it is the granite that can be carved so dramatically by glaciers, the granite areas, such as the sharp crests from Mount Whitney to Yosemite Valley, the Castle Crags, and the Trinity Alps, are breathtaking.

Today Sierran climate differs also from that of the Klamaths. Although the Klamaths do receive snow, temperatures are fairly mild in winter, so that they are not generally snow clad. Nevertheless, the area is quite wet; Monumental (near the Oregon border) averaged almost 28,000 mm (nearly 110 inches) annually in a recent six-year period. Most of this precipitation is rain, not snow, and therefore contributes to rapid runoff rather than spring snow melt. Because of this high rainfall, and because of the swift rivers in these steep mountains, the Klamath Mountains are eroding faster than any other place in the United States. However, part of the record-breaking erosional rate in the cut-over area adjacent to Redwood National Park is due to unwise lumbering practices.

Northeastern California, in contrast, is a land of volcanoes. Two of the highest are still active: Mount Shasta, 4,317 meters (14,162 feet) and Mount Lassen, 3,187 meters (10,457 feet). They are members of the Cascade Range, which extends northward to the Canadian border.

The northernmost peak in the Cascade Range is British Columbia's Mount Garibaldi, 2,678 meters (8,787 feet); the southernmost peak is generally considered to be Mount Lassen. Lassen erupted in 1914–1917, the only volcano in the conterminous forty-eight states to erupt in this century, until 1980, when Mount St. Helens, Washington, erupted spectacularly. In the Cascades are several volcanoes that have been active very recently, geologically speaking. Almost all these Cascade volcanoes lie near the Columbia Lava Plateau, or project from it. In California, the two major cones, Mount Lassen and Mount Shasta, grade into the Columbia Lava Plateau (called the Modoc Lava Plateau in California).

The Modoc portion of the Columbia Plateau, covering the northeast corner of California, is about 26,000 square kilometers (10,000 square miles) in size, or about one-twentieth of the entire plateau, which extends into Oregon, Washington, Nevada, Utah, and western Wyoming. The Modoc Plateau today is a rolling upland, dotted with volcanic cones and flows. Despite its fiery origin, the region is now one of long, cold winters. Precipitation, however, is light: Yreka has an average annual precipitation of only 457 mm (18 inches), and Alturas is semiarid, averaging only 330 mm (13 inches) of rain each year.

South of northeastern California and the Klamath Mountains, and between the Sierra Nevada and the Coast Ranges, lies California's agricultural treasure chest, the Great Valley. The northern part, drained by the Sacramento River system, is commonly called the Sacramento Valley; the southern part, drained by the San Joaquin and its tributaries, is called the San Joaquin Valley. The two join at the 1,900-square-kilometer (750-square-mile) "delta," which has been cut by the two braided rivers into many islands, providing more than 1,600 kilometers (1,000 miles) of waterways.

The flatness of the Great Valley was derived through its geologic history. When the Sierra Nevada, Klamath, and Peninsular ranges rose from the sea, the edge of the Great Valley became the new coastline. Gradually, as the Sierra was worn away, the area where the valley is today began to be filled with debris from the mountains, and when the new Coast Ranges rose on the west, the sea between the Coast Ranges and the Sierra dwindled even more. By about 10 million years ago, most of today's valley area had become dry land. During the Great Ice Age that carved the Sierra Nevada into its present form, a huge freshwater lake occupied much of the valley. Gradually, it, too, dried; but dur-

FIG. 5. Sutter Buttes.

ing heavy storms even in historic time, the valley has resembled a vast inland sea.

About three million years ago, a small volcano rose in the Great Valley, culminating in explosive eruptions during the Great Ice Age. Today, this volcano, known as the Sutter or Marysville Buttes, is the highest point in the Great Valley, standing 650 meters (2,132 feet) above sea level. Although it is separated by many miles from the nearest active volcano, Mount Lassen in the Cascade Range, it may be the most southerly volcano in that range.

The Sacramento Valley has a warm climate, with cool, moist winters and clear, hot summers. July temperatures at Sacramento average 24°C (75°F); average temperatures at Redding are higher at 28°C (83°F). Precipitation ranges from 457 mm (18 inches) at Sacramento to 965 mm (38 inches) at Redding.

The San Joaquin Valley is drier and hotter. The winters are cool, but not as moist. Precipitation ranges from an average of 356 mm (14 inches) at Stockton to only 152 mm (6 inches) at Bakersfield. Very seldom does snow fall. Summers are hot; many summer days in July and August are over 38°C (100°F).

To the west of the Great Valley lie three provinces that border the Pacific Ocean—the Peninsular, Transverse, and Coast ranges. Of these, the Coast Ranges are the longest and wettest. About 965 kilometers (600 miles) long, north to south, the province extends into Oregon on the north, abuts against the Transverse Ranges on the south, and hems in the Great Valley to the east.

The Coast Ranges are not high. Mountain crests average between 600 and 800 meters (2,000 to 4,000 feet); a few peaks exceed 1,800 meters (6,000 feet). Two prominent mountains near San Francisco, Mount Tamalpais, 794 meters (2,604 feet) and Mount Diablo, 1,173 meters (3,849 feet), rise from sea level and seem lofty in contrast to their surroundings. Because the Coast Ranges were not high during the Great Ice Age, they were not glaciated; for that reason, California does not have glacially formed fjords as does Alaska and other regions farther north.

The climate of the Coast Ranges is of the Mediterranean type, with relatively cool summers and mild winters. Rarely does the temperature drop below freezing, and rarely does it exceed 30°C (100°F). In San Francisco, January temperature averages 8.5°C (51°F), while the hottest month, September, averages 13°C (62°F). The area receives an average of 533 mm (21 inches) of rain per year. Mediterranean climates are rare; less than 1 percent of the Earth, and no other area in the United States has such a climate.

A glance at a relief map of California will show that the Transverse Ranges, south of the Coast Ranges, have a trend, or "grain," opposite to that of other mountains in the West. They are high mountains; the highest point, Mount San Gorgonio, reaches 3,506 meters (11,499 feet). The province not only borders on the sea but plunges beneath it to the west, leaving only the highest peaks of its western part to poke above the ocean as the Channel Islands.

The Transverse Ranges province extends from the pleasant coast of Santa Barbara eastward to Joshua Tree National Monument, where it merges with the Mojave Desert. It is bounded on the north for much of its length by the San Andreas fault system. Both the San Andreas and the province trend eastward here; what caused mountains and fault to have this unusual alignment is not yet fully understood. There are few mountain ranges in North America with this trend.

Like the Coast Ranges, mountains in the Transverse

Ranges province are still rising, leaving faulted, oil-rich basins between. And the Transverse Ranges are rich: rich in mineral and energy resources (more than 700 million metric tons of oil have been produced) and rich in rocks and fossils. There is a wider spectrum of rocks of diverse ages here than elsewhere in the state, as well as the world's thickest section of Pliocene sediments, an aggregate of 3,650–4,250 meters (12,000–14,000 feet) containing fossils of animals that lived in seas of that time (7–3 million years ago).

Within the Transverse Ranges, in the heart of the city of Los Angeles, lies Rancho La Brea, the world's most famous fossil locality. Here were and still are "tar" pits—asphalt pools that formed by oil oozing upward from the rich petroleum deposits below. As the pools were often covered by a film of salty water, they lured many birds and animals that lived during the Great Ice Age to their sticky shores. Many of the creatures were trapped, and predators, enticed by their frantic cries, in turn met death when they attempted to attack.

Altogether, nearly 50 species of mammals, 110 species of birds, a few kinds of snakes, turtles and toads were caught, as well as many insects and other invertebrates. Fossils of plants that were also preserved show that pine, cypress, ash, and manzanita grew on the shores of the tarry lake. Local human inhabitants must have known of the danger, for human fossils, though present, are scarce.

The Transverse Ranges are rich also in natural and man-induced geologic problems, particularly in the type that once were labeled "acts of God"—that is, earth slides of various kinds and earthquakes. Earthquakes, common in the region, are an indication that these mountains, like the Coast Ranges, are still rising.

Southern California's climate is generally mild, having warm almost rainless summers and pleasant winters. Santa Barbara, on the coast, receives 457 mm (18 inches) of rain in the winter, making it much sought after as a place to live, as is the rest of the Transverse Ranges, including the densely populated Los Angeles Basin.

Although the Sierra Nevada is the longest continuous mountain range in the conterminous United States, the Peninsular Ranges, which extend from the boundary of the Transverse Ranges southward beyond the Mexican border to the tip of Baja California are longer, having a total length of 1,250 kilometers (775 miles). On the east, they are bounded by the Colorado Desert and the Gulf of California; on the west, they extend offshore to the drowned edge of the continent, making a strip varying in width from 50 to 60 kilometers (30 to 100 miles).

Geologically, the Peninsular Ranges are made up of igneous rocks of about the same age as those in the Sierra Nevada, as well as metamorphic rocks related to those of the Sierra. Although the Klamath, Sierra Nevada, and Peninsular provinces are called by different names, there seems little doubt they should be considered as one long mountain range, broken in the south by faults, and covered in the north by overlapping lava flows.

The mountains of the Peninsular Ranges seem unusually high, as they rise boldly above low-lying valleys. Mount San Jacinto, highest peak of the range, forms a towering backdrop 3,293 meters (10,805 feet) above the town of Palm Springs. Most ranges that make up the province are miniature Sierras in shape, with sharp eastern faces and gentle western ones.

The Peninsular Ranges have been a highland undergoing erosion for the past 100 million years. Erosion has unearthed treasure in them, as it has in the Sierra Nevada, but little of it was gold. Gemstones and unusual minerals have been the Peninsular Ranges' most intriguing contributions. Many faults lace the province, which has had numerous earthquakes.

Along the shoreline near San Diego, a series of terraces is preserved that marks the rising and falling of the sea.

Some of the changes in sea level were due to melting or freezing of the great glaciers; others mark changes in elevation of the Peninsular mountains by earth movements.

The climate of the Peninsular Ranges is dry, as shown by the sparsely treed mountains. San Diego, on the coast, receives an average of about 250 mm (10 inches) of rain each year. Snow anywhere in the southland is scarce except in high mountains; in San Diego it is virtually unknown.

California's desert country, as it is usually thought of, consists of three geomorphic provinces: the Mojave Desert, the Colorado Desert, and the Great Basin. The Mojave is the "high desert" of weather reports; the Colorado is "low desert." The Great Basin province, in California, is steppe, or "high plateau," although it contains Death Valley, the nation's lowest lying desert. Throughout much of their geologic history, the three provinces were part of the Great Basin, and may still be considered so from a structural point of view.

The Mojave portion of this desert area has been separated from the rest by movement along the San Andreas and Garlock faults. From their junction, near the southern end of the Sierra Nevada, the two faults spread out like handles of a nutcracker. Between lies the Mojave Desert.

Many other faults show on maps of the desert. The Great Basin is sometimes called "Basin Ranges" or "Basin-and-Range" province, indicating that mountain ranges in it alternate with basins. Most of the ranges are bounded on the east and west by faults.

Faults in the desert have moved rock masses long distances. One rock sequence in the Orocopia Range, north of the Salton Sea, is nearly identical to a sequence in the San Gabriel Mountains of the Transverse Ranges on the opposite side of the San Andreas fault, 160 kilometers (100 miles) away.

Granitic rocks are fairly common in the mountains of the California deserts (four "Granite Ranges" are named on topographic maps in the Mojave Desert alone), a relic of the history the Californa desert has shared with the Sierra Nevada and other ranges of the Great Basin.

Travelers in the desert are particularly aware of volcanoes, lava flows, and dry lakes, all especially visible in the dry desert air. Volcanoes and their products are not dependent upon climate, but they persist longer as landforms where water for weathering is scarce. Dry lakes, however, are a result of desert climatic conditions.

Today's desert climate can be blisteringly hot (56.7°C— 134°F has been registered in Death Valley) and dry (Death Valley receives less than 50 mm (2 inches) of rain in an average year), but parts of the high desert can be quite cold in winter. The crest of the White Mountains, the next range east of the Sierra Nevada, is so high, cold, and dry that it qualifies as a "cold" desert, in company with Antarctica and parts of Alaska.

A part of California that many of us might not have considered as such is the strip that lies offshore. How far California extends out to sea is a political question that is not entirely settled, but this offshore portion is our new frontier. It is a province with geology and topography, but instead of rain and sun, the ambient sea is its weather. Life and death go on within it, as on land; its floor is being eroded, and new rocks and sediment are being added. We know little about it as yet, but what we do know intrigues us.

"Nature in these old hearths and firesides has given us beauty for ashes."

4 · VOLCANOES

Volcanic eruptions have produced the most elegantly symmetrical and delicately colored features in the California landscape. Ancient volcanoes were active as far back as we are able to trace California's geologic history, and today's volcanoes could erupt at any time. In this long volcanic history, much of the record has been changed so many times that it is not always easy to tell which pages were originally volcanic. Erasures by metamorphism and erosion have also blurred the story.

However, there is no dearth of volcanic features produced within the last 10 million years, many of which are so fresh that they look as if the eruption were yesterday.

Among the freshest of California's volcanic features are small "cinder cones" built up by hot volcanic sand spewing from the crest. Rarely more than a few hundred feet high, these beautifully symmetrical, miniature volcanoes dot the southern desert and the northern volcanic country. Such cones grow quickly, even in human terms, reaching their heights often in a matter of days or weeks. While they are active, lava fountains play from vents at their crests, building up cinder cones in bursts of spectacular pyrotechnics.

When the lava has ceased to play, the cooling cone is left, looking like a pile of hot, black sand. Here and there, larger pieces that have been cannonaded from the top dot the landscape. Called "bombs," these rounded and twisted rock masses are blebs of fluid lava that cooled quickly while hurtling through the air. The main mass of the cones, however, is made up of fine volcanic particles the size of sand that originally were drops of lava.

The world's most famous cinder cone is Paricutín, Mexico, which rose from a flat cornfield to the surprise of the family farming it. The cone reached 8 meters (25 feet) by the second day of its life, 125 meters (400 feet) within a week, and by the end of a month had reached its 300 meter (1,000 foot) fairly steady height. It was more or less continuously active for 9 years, from 1943 to 1952; tourists came from all over the world to watch its flashy displays and to climb the cone's nearly perfect, 30° slopes.

Cinder cones are rarely higher than this, but they are the steepest of volcanic cones. As they are made of sand-sized material, 30° is their normal grade—similar to the steep slope of a sand dune. Other cones are flatter: composite volcanoes (also called "stratovolcanoes") form a much more obtuse slope angle, and the huge shield volcanoes are quite flat, some with slope angles less than 1 degree, although 1 to 10 degrees is common. Because we see the beauty of symmetry in volcanoes, our eyes tend to exaggerate the steepness.

The only one of California's many cinder cones to erupt in historic time is Cinder Cone in Lassen Volcanic National Park (fig. 6). Doubtless many other young volcanic cones have erupted while people have been here, but we have no written records of them. Cinder Cone, however, lay along the path of William H. Noble's immigrant trail. For the then great sum of $2,000, he would reveal the route to parties of immigrants. The trail was well used; whether the travelers were bona fide toll payers or freeloaders, enough wagon trains passed over it to wear it down so that it is still visible today.

Travelers on the trail as well as scientists saw Cinder Cone erupt in 1851. By then, the 210-meter (700-foot) cone had already long been built; its major structure was about

FIG. 6. Cinder cone, Lassen Volcanic National Park, as it appeared in 1895. The cone last erupted in the 1850s.

500 years old. A hiking trail leads up the cone today, passing by a basalt lava flow that issued from the side, and culminates on the steep rim. From the edge of the rim, one can see the area covered by ash, cinders, bombs, and lava. After the cone was built, the flow of blocky basaltic lava one crosses on the trail flowed out toward Lassen Peak and was covered by more ash and cinders before it was completely cool. Where the ash fell on the hot lava, rising steam oxidized it to red and yellow; hence the name, "Painted Dunes."

The spurt of activity the immigrants saw in 1851 was the latest in the cone's history, but was not necessarily its last. It did not erupt when Lassen Peak did in 1914.

Cinder cones do not last long where there are rains to erode them. Soon, in wetter climates, a pattern of rivulets is worn down the side of a cone. As rain and snow are commonly blown by storms from one dominant direction, often a cone is cut apart on that side, to be dissected more thoroughly as time goes by. Many of California's cinder cones are in desert regions with little moisture, and for that reason, cones that are thousands of years old look fresh and new.

Some of the cinder cones one sees are red, like the Painted Dunes. Along U.S. Highway 395, a twenty-two-thousand-year-old cinder cone near Little Lake shines red in the afternoon sun. The cone is truly red; it is not merely the fading light of afternoon sunset that gives it color. The redness of such cones, however, is acquired. The tiny lava droplets, which may have been red hot when spewed from the crater of a cone, were black when they cooled, but the iron in them has since rusted (oxidized) to red. The particles are called "cinders" because they resemble the cinders removed from coal-burning furnaces.

The pastel colors of other volcanic beds owe their shades to chemistry, as well. Generally, oxidation of iron is responsible for the pink, tan, mauve, lavender, and pale blue of the hillside rocks, but manganese and other elements may also play a part. Artist's Palette, Death Valley, is a good example of the many colors that volcanic rocks can display.

Largest and most spectacular of California's volcanic forms are its towering volcanoes, of which several are still classed as active, including Mount Shasta, Mount Lassen, and Glass Mountain. Some, like Mount Konocti on Clear Lake and Mammoth Mountain on the east side of the Sierra Nevada, have been active in the geologically recent past and could at any time exhibit renewed vigor. Although Mammoth Mountain reached its greatest height 370,000 years ago, hot springs on its side show that the mass is not yet wholly cool.

Mightiest of the active volcanoes is Mount Shasta, 4,319 meters (14,162 feet) high, which towers 3,000 meters (10,000 feet) above the surrounding land like a sentinel (fig. 7). Shasta is near the south end of the Cascade chain, which extends north into Canada.

Mount Shasta is classed as a "stratovolcano"; that is,

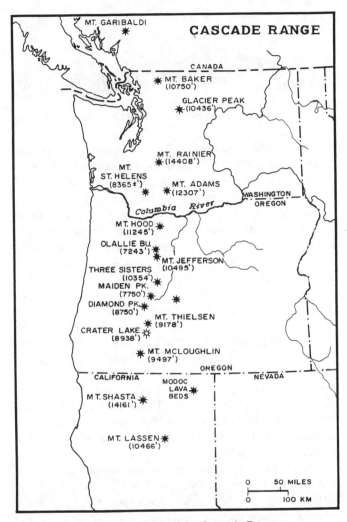

MAP 4. Volcanoes of the Cascade Range.

FIG. 7. Panorama of the Mount Shasta area.

one composed of alternating layers of poured out lava and hot fragments ("tephra") thrown out of its several craters. It is a massive old structure, with a total volume of 330 cubic kilometers (80 cubic miles) of tephra and lava.

In 1786, the French explorer Comte Jean François de Galaup de La Perouse, sailing along the Pacific coast, saw what he thought was a volcano erupting on land. Which volcano that was has never been identified with certainty. It may have been Mount Shasta.

By 1786, Shasta was as tall as it is today, having achieved that eminence in a four-stage history. In its first volcanic stage, Shasta formed the Sargent's Ridge cone, on its south flank (table 3). This cone is more than 100,000 years old, as indicated by its relationship to nearby glacial deposits. Next to be formed was the Misery Hill cone, near the present crest. It is not as old as Sargent's Ridge, but is more than 12,000 years of age. This is the part of the mountain one sees from the ski lift. The pumice of Red Banks, immediately above the lift, was deposited during the last episode of this stage.

Third event in Shasta's history was the development of the smaller daughter cone, Shastina (see table 3). From some vantage points, Shastina is not visible; from others, it is so prominent that the mountain looks as if it were twin peaked. Shastina's age is better known than that of the other cones. During the last eruptions of Shastina, hot avalanches poured from the summit. These have been dated by radiometric methods as being 9,400 ± 200 years old.

Last event in Mount Shasta's volcanic story to date was

the eruption from the present vent, which poured out lava flows beneath the glaciers that mantle the mountaintop. Since these glaciers are products of the "Neoglacial Age" or "Little Ice Age" of the past 10,000 years and are not relics of the Great Ice Age, the lava must be younger than they are. Some lava flows may be less than 1,000 years old; in fact, radiometric dating indicates that some may be less than 200 years old. If so, Shasta may well have been the mountain La Perouse saw.

It is possible that extraordinary fumarolic or hot spring activity caused some of the glacial mudflows that have cascaded down Mount Shasta in historic time. The Icelandic word "jökulhlaup" is used to describe such mudflows, because Iceland has several volcanoes that sometimes erupt beneath glacial ice, giving Icelanders familiarity with the phenomenon. In California, however, glaciers are minute and most volcanoes are quiescent, so that Shasta is virtually the only site where such a mudflow is likely.

Warm weather, not volcanic activity, was officially

TABLE 3. Major Cones on Mount Shasta

Major cone	Central vent	Principal flank vents	Estimated age (years)
Hotlum	Mount Shasta summit	None	<4,500–190
Shastina	Shastina summit	Black Butte	<12,000–9,400
Misery Hill	Misery Hill saddle	Near Gray Butte	<100,000–>12,000
Sargent's Ridge	Head of Mud Creek	Red Butte, near Gray Butte, McKenzie Butte, North Gate, northwest flank of Spring Hill	>100,000

blamed for the last glacial breakup on Mount Shasta, but hot spring as well as volcanic activity may have triggered some of the previous jökulhlaups that have occurred in the past 1,200 years. Traces of prehistoric mudflows date back to A.D. 800, and in the past century there were mudflows in 1881, 1920, 1924, and 1931.

Because the mudflows consist only of glacial flour, boulders, and water, with no organic components, the U.S. Forest Service has set aside parts of them for the study of soil formation and revegetation. No wood cutting (not even dead trees), timber or Christmas tree harvest, or grazing is permitted. Although existing roads may be used, off-road vehicles are not permitted. Such vehicles, which can compact acres of future soil in a single pass, can easily destroy the area's potential as a study laboratory by slowing down or preventing soil formation.

Mount Shasta still has active fumaroles (steam vents) and an acid hot spring that show that the summit is still cooling.

Mount Lassen is not as symmetrical as Mount Shasta. It is a dome, rather than a cone, and is surrounded by heaps of broken rock (talus). Mount Lassen is the southernmost of the Cascade volcanoes, and was the only one of the entire range that had erupted in this century until Mount St. Helens burst out in 1980. Mount Baker, in Washington, made threatening gestures in 1975–76, but nothing had come of it by 1982.

Lassen Peak (fig. 8) is a 3,186-meter (10,453-foot) volcano standing about 130 kilometers (80 miles) south-southeast of Mount Shasta. Like Shasta, it was constructed about 10 million years ago when an ancestral volcano, Brokeoff, rose to a height of about 3,350 meters (11,000 feet). Brokeoff collapsed; remnants of it may still be seen in Brokeoff cone, just outside of the park. When it was at its largest, this old volcano had a diameter of 15 miles. The lavas of today's Lassen began to erupt from its northeastern side during the Great (Pleistocene) Ice Age. Volcanic ac-

FIG. 8. Mount Lassen.

tivity has continued in and near Mount Lassen, forming Cinder Cone, Chaos Crags (about A.D. 500), and other volcanic features, but there was no historic record of eruption of Lassen until 1914, except for wisps of steam that were seen rising from the top in 1857.

Mount Lassen made up for its quietness by giving a spectacular show from 1914 to 1917. On May 30, 1914, the summit began to erupt; in the next two years watchers counted 150 distinct eruptions. On May 19, 1915, a huge volcanic mudflow (lahar), deriving its water by melting snow, roared down the northeastern side of the mountain, covering parts of ranches and barely allowing the few ranchers to escape. The seething mudflow was thick enough to buoy up large rocks, which it carried many miles down Hat and Lost creeks.

Three days later, under cover of darkness or under the cover of a huge mushroom cloud that soared upward 10 kilometers (5 miles), a glowing cloud (*nuée ardente*) sped down the same side of the mountain, attaining a speed of perhaps as much as 200 kilometers (100 miles) per hour. Fortunately, those people who might have been in the way

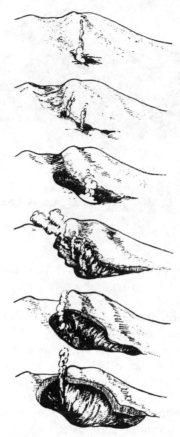

FIG. 9. Lassen's explosion crater as it enlarged progressively during 1914. These drawings were made from photographs of the crater taken by some climbers who were on the mountain during its eruption, and from available measurements and descriptions. All photographs were projected by the artist so that the point of view is the same for all drawings. The entire 1914 crater was filled with lava on May 18, 1915. Two smaller craters developed in 1915, and a larger one in a final burst of effort in 1917. All three are marked and visible in the crest of the peak today.

FIG. 10. Five stages in the destruction of the lookout atop Lassen Park, 1914. *Top*, the lookout as it was before the eruption; *second from top*, the roof of the cabin had been punctured by rocks; *center*, further demolition of the cabin before its destruction; *next to bottom*, cabin remains on October 20, 1914, after 72 eruptions; and *bottom*, only a pole remained to mark the site.

FIG. 11. The 1915 mudflow on Mount Lassen as seen on May 24, 1915.

FIG. 12. The mushroom cloud formed in the largest of Mount Lassen's eruptions.

of it had moved out of its path because of the previous mudflow. There was a forest in its path, however, which could not escape. Remnants of it may still be seen in the "Devastated Area." (The Caribbean volcano, Pelée, produced this same type of hot blast in 1902, killing all 38,000 inhabitants of the town of St. Pierre, except for two, one a prisoner protected by his airless cell.)

This was the volcano's *piéce de resistance*; activity has waned from that time, until today only wisps of steam, boiling mud pots, and solfataras remain to remind us that there is life in the old dome yet.

The solfataras, or sulphurous fumaroles, particularly well developed at the Sulphur Works, are further reminders that a volcano produces a large amount of gas, which is the chief cause of the explosions. The principal gas is, of course, steam. The boiling mud pots, derived from hot springs, as well as the hot springs and steaming fumaroles, are clues that there is still heat in the mountain. Volcanoes tap a source of heat that has rendered the underlying rock fluid (underground fluid rock is called "magma"). Volcanologists suppose that at some depth beneath the earth near volcanoes there are "magma chambers" where reservoirs of molten rock lie. These hot bodies are the source of both heat for the hot springs and lava for the volcano.

Among California's other large volcanoes, Glass Mountain and Little Glass Mountain in Siskiyou and Modoc counties are classed as active. Mammoth Mountain, in the Sierra Nevada, erupted 370,000 years ago but still steams now and then; Mount Konocti, Lake County, was built between 400,000 and 250,000 years ago during a series of ten or more eruptions. At one point in the construction of Mount Konocti, two cinder cones shot lava fountains high into the air; later, these were partially buried by fluid lava.

The nation's first geothermal energy field is located southwest of Clear Lake at The Geysers. Here, electricity is being produced from steam heated by the Earth itself. The

FIG. 13. Mount Konocti.

heat source is probably a body of uncooled magma deep within the Earth that could furnish lava for a new eruption.

Volcanoes like Mounts Lassen, Shasta, and Pelée that explode reverberatingly, throwing mushroom clouds into the air, are dominantly "andesitic" volcanoes. They spew out lava and volcanic particles intermediate in chemical composition between the dark, basaltic lava (such as Hawaiian volcanoes produce) and very light-colored lava. Andesite is a gray, solidified lava, too fine grained to reveal its mineral composition by casual inspection.

In general, dark, black basaltic lava is hotter and more fluid than grayer andesite, and even more fluid than cooler, light-colored, stickier rhyolite. Although a volcano may produce all three types of lava during its lifetime, it generally produces predominantly one kind.

The slopes of Lassen and Shasta are covered with andesite; many eruptions of both peaks produced layers of andesite and a relative, dacite, now visible as distinct layers on the mountainside. It was a volcanic mud flow made of andesite that preserved Herculaneum, twin city to Pompeii, when Vesuvius erupted in A.D. 79. Because the flow arrived slowly, most inhabitants had time to flee; because it hardened into tough rock, the city is only now being slowly

excavated. But everyone did not escape. In 1982, archae-
ologists disinterred eighty skeletons, including one woman
with rings on her finger bones.

Many similar mudflows starting from various vents near
the mountain crest poured down the slopes of the Sierra
Nevada beginning about 20 million years ago. "Lahar" is
the term used by volcanologists for such volcanic mudflows. It is an Indonesian term, for such flows are common
in Indonesia.

In the northern Sierra Nevada, lava, mixed with winter
snow while still hot and fluid, was propelled downhill as a
syrupy mud, jumbling together pieces of hardened lava,
rocks torn from the mountainsides, and ripped and shat-
tered trees, all mixing and churning in a steaming paste. It
happened over and over again, piling layer upon layer. By
the time the andesitic episode ended, 15 million years later,
more than 31,000 square kilometers (12,000 square miles)
of land surface was covered, and more than 8,000 cubic
kilometers (2,000 cubic miles) of new rock had been added
to the mountains!

Andesitic Sierran terrain of today retains the look of
chaos. The old mudflows are now grass-covered, tree-
studded, rolling, lumpy landscapes dotted with boulders
rounded in their exciting trip downhill. Here and there the
flows have been worn to castellated badlands, turrets, and
ghostly cities.

All of the volcanic material that is exploded from a volcano is called by specialists "pyroclastic," from Greek
roots meaning "fragments from the fire." Sometimes the
noun "tephra," another Greek word (meaning "ashes"),
first applied by Aristotle to volcanic material, is used in
place of the longer adjective-noun combinations (pyroclastic rock, pyroclastic material, or volcanic ejecta).

Smallest of the tephra is dust; ash is somewhat coarser,
but still very fine; cinders are small fragments; lapilli (Latin
for "small stone") and scoria are pebble sized; while blocks
and bombs are the size of boulders and larger.

MAP 5. California's volcanic country.

Volcanoes may bubble over, rather than exploding, pouring out froth similar to froth from boiled over applesauce or candy. The volcanic froth, cooling quickly in air, traps gas in the multitude of bubbles within it. For this reason, pumice (which is congealed volcanic froth), will often float on water—not a common attribute of rocks.

Or if the liquid itself cools very quickly, the resulting

rock is a natural glass, of which black obsidian is the most common. If you have ever made clear hard candy such as that in all-day suckers, you will recognize the froth you must scoop off (pumice) and the clear, quickly cooled candy (obsidian glass). One may find natural glass and pumice near many California volcanoes, but an unusually good place to see a small hill of glass is at Obsidian Butte on Glass Creek, Mono County.

Andesitic volcanoes can be explosive and dangerous, but they are not, on the whole, as explosive or dangerous as those that spew out lighter colored, rhyolitic lava. Rhyolite lava, like andesite, is fine grained, but its chemistry indicates that it is about three-fourths silica (SiO_2), while andesite is about two-thirds silica and basalt is about half. On the whole rhyolite is a cooler lava, less fluid than others. In some places, rhyolite has formed tough, flat flows; in other places, it pushed upward as a sticky mass, while in yet another setting it exploded with devastating violence.

Flat rhyolite flows are not as common in California as other flows, or as other forms of rhyolite. Rounded lumps of rhyolite called "craters" (a misnomer) or domes, form chains in the landscape east of the Sierra Nevada, in the Basin Ranges province. The domes rose, one at a time, as sticky masses of rhyolite, finally protruding from the ground much as toothpaste from a tube. As the mass reached the air, it cooled quickly, cracking open. Mono Craters (fig. 14), guardian of Mono Lake, is a chain of rhyolite domes. Inyo Craters, nearby, is another; Coso Domes, to the south, is less a chain than a dome field. The oldest of the Mono Craters is about 60,000 years of age; the youngest is about 1,300 years old. Most of the craters contain some obsidian (volcanic glass). In the same area are other smaller domes, such as Obsidian Butte which is made entirely of obsidian.

South of Mono Craters, travelers on U.S. Highway 395 pass another group of rounded domes—the Coso Domes. One, called "Sugarloaf," is easily seen from the highway. Nearly every state has its "Sugarloaf Mountain"; some

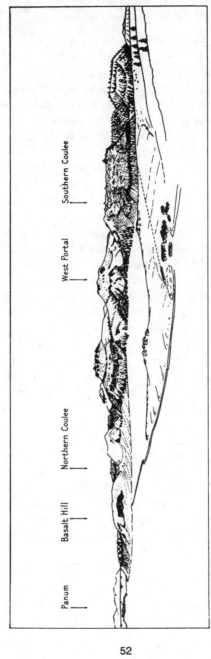

Panum   Basalt Hill   Northern Coulee   West Portal   Southern Coulee

FIG. 14. Panorama of Mono Craters.

52

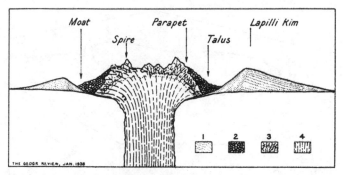

FIG. 15. Structure of a typical obsidian dome: The outer rim, formed of fine ash and small volcanic fragments (1), is marked by dots; the core of the dome is made of black, glassy obsidian (4). The more frothy top of the dome may be brown obsidian (3). After the dome has risen, it begins to weather, and blocks break off to form heaps of rubble (2).

have several. The word "sugarloaf" is a puzzle to us now, but in the nineteenth century, when many American mountains were named, sugar was sold by the "loaf." Rounded loaves were a common sight in general stores, and when grocery windows became "modern," the sugarloaf was a principal decoration.

Sugarloaf-shaped domes, being more rounded, do not give the impression of steepness that other volcanoes do, but this may be deceptive; their lower slopes can be quite steep.

Upside-down sugarloaves may be left in volcanic country when volcanoes blow out; one looks down into a bowl at these volcanoes, rather than up at a loaf to see layers of ash and other volcanic material lining the crater walls. In wet country, a lake may form in the bottom of a blowout, as it has at the 1,500-year-old pumice cones called Inyo Craters, on the east side of the Sierra Nevada; such lakes are called "maars." The two Ubehebe craters in Death Valley National Monument are blowouts with low rings of volcanic ash and lapilli around them, but no lake stands in them.

Stubby domes like Mono Craters and Coso Domes were

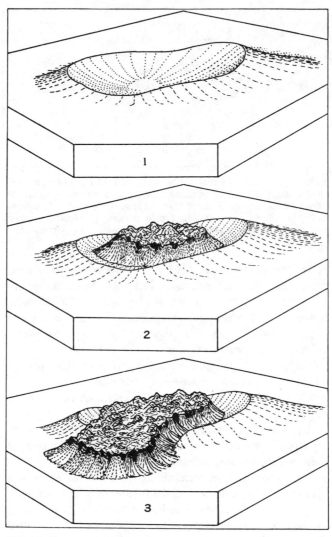

FIG. 16. How an obsidian dome develops. 1. A low cone has formed by explosions of pumice. 2. A volcanic dome rises in the center of the dome, pushing upward. 3. The dome has risen so high that part of it has pushed over the crater rim.

not very dangerous as they formed because they did not explode. It may be that there were once other domes in the California landscape that were blown apart by explosion of gases inside them—one cannot now tell. The remains of such explosions can be seen, but whether or not a dome formed before the big bang we do not know.

A huge explosion shook Long Valley, near Bishop, about 700,000 years ago. A crater, perhaps Long Valley itself or Glass Mountain, blew up, throwing hot lava fragments and ash high into the air. Ash from this explosion has been found as far away as Utah. Volcanic fragments, many still white hot, dropped in a "rain of fire" over the landscape. In a very short time, humanly and geologically speaking, 150 cubic kilometers (35 cubic miles) of incandescent fragments filled the valley with a blanket of ash perhaps 300 meters (1,000 feet) thick. Fresh fragments of this type of ash look much like children's jacks, but settle together into a tight mesh as they cool.

When it first dropped, the temperature of the ash must have been at least 650°C (1,200°F), but it eventually cooled and settled to half its original thickness. As the sheet cooled, plumes of steam rose through vents from the deeper, still hot parts. Those vents (called "fumaroles") remain today, looking like chimneys on the otherwise fairly smooth surface of the ash bed.

The incandescent volcanic ash was cooled into a rock called "tuff"—a chemically resistant, but soft rock. Prehistoric residents used the tablelike flow (which diverted the course of the Owens River) as billboards; several groups of petroglyphs are carved on smooth exposures near Bishop. What the petroglyphs meant is not well understood, but today they are visited by tourists who view them on a marked trail called the "Petroglyph Loop."

The softness of the tuff made it easy for rock artists to leave their messages for the world. Because tuff is soft, it is easily weathered by wind and rain. Tuff beds in California have not been as dramatically weathered as the Bandelier

Tuff in New Mexico, where the forces of weather have etched the light-colored rock into a honeycomb, which was further modified by residents during the eleventh and twelfth centuries into a streamside apartment house. Honeycomb tuff in Anatolia, Turkey, has been carved into dwellings that have been in use for 5,000 years. Some of the cave residences have seven underground floors.

Hard lava is not as easily cut into dwellings as tuff; nevertheless, all types of lava have been cut into building stone blocks for use in masonry construction. Basalt quarries in the Coast Ranges furnished stone blocks for San Francisco's streets in the last century; today these hard blocks are being recycled as prized garden stone.

Basaltic eruptions are, on the whole, less dangerous to observe than rhyolitic or andesitic ones. This is because the lava is less explosive. It is also darker, higher in iron and magnesium, and more fluid; for these reasons, it tends to flow downslope rather than explode. Scientists accustomed to working around the erupting Hawaiian volcanoes learn that they can, with care, step on the cooled crust while beneath, the red hot lava seethes. Since it is reasonably safe to work near these basaltic, frequently erupting cones, the U.S. Geological Survey has set up a laboratory on the slopes of Kilauea that has yielded a great deal of valuable information about the mechanics of volcanic eruptions.

Scientists have learned, for example, how the flows move; how lava tubes are formed; and how basaltic "pillows" acquire their rounded shape.

The scientists watched the flows in Hawaii moving as rolling, tumbling masses, swallowing already cooled pieces of their own roofs and sides back into their main moving flood. Eventually, the flows halt, leaving rough-and-tumble lava surfaces (called by their Hawaiian name, aa) behind. Most California flows are of this aa type (pronounced "ah-ah" as if you were having your throat inspected), but a few are of the ropy type called "pahoehoe," which forms if lava halts in a lava lake while still fluid.

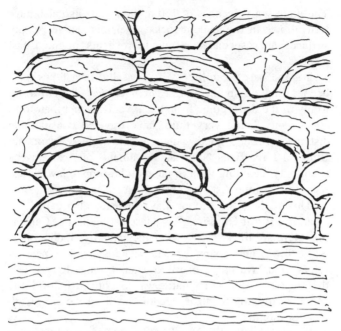

FIG. 17. Pillow lava, formed when hot lava erupts in or flows into water.

In places where volcanic flows can travel downhill to the sea (as in Hawaii or Iceland), the flows send up vast jets of steam when they strike the sea. Hot lava continues to flow underwater, eventually cooling there. In the same way as lava that originates from undersea volcanoes, it may form a mass of interconnecting rounded "pillow" shapes. Geologists guessed that these strange pillows found in the rock record were formed underwater; the guess was proven correct when diving geologists photographed a cooling lava flow in the process of forming pillows on the bottom of the shallow sea.

In California, a lava flow formed pillows when the lava ran into Golden Trout Creek, in the high Sierra; another made pillows in the San Joaquin River at Devils Postpile.

The undersea volcanoes of 150 million years ago also have left pillows in the ancient rocks. Their distinctive rumply surfaces are especially visible in roadcuts in the Coast Ranges.

For the past 30 million years, basaltic lava flows have been piling up in the northeastern corner of California. Many of the flows come from linear vents, rather than from cones, and, being very fluid, have traveled considerable distances. From a good vantage point, such as Schonchin Butte in Lava Beds National Monument, one can have a seemingly unending vista of faulted, flat-topped lava flows, receding into the distance like a giant staircase.

This region, the Modoc Lava Plateau, also called the "Land of Burnt-Out Fires," is a fascinating part of California. Here are thick lava flows, punctuated by cinder cones, glass flows, and young volcanoes; here are disappearing rivers, and lava caves many kilometers long. It was here that the last major Indian battle was fought in 1872–73: the Modocs, fighting for their freedom, held off the U.S. Cavalry for months.

In places the disappearing rivers drop into the lava tubes—long, thin caves within hardened lava flows. Lava tubes are fairly ephemeral as landscape forms, rarely persisting more than 100,000 years before being filled or obscured. For that reason, the world's lava tubes are in young volcanic rocks, dating from the last part of the Great Ice Age to the present day. Lava tubes are formed only in basalt; other volcanic rocks are too explosive. Also, they form only in the ropy pahoehoe type of basaltic lava, which is hotter and more fluid than the chunky, churned-over blocky aa.

Lava tubes are scattered throughout the volcanic country of the western United States, as well as Mexico, the Leeward and Azores islands, Iceland, Japan, New Hebrides, Australia, and Kenya. In the western United States, Alaska, Arizona, California, Hawaii, Idaho, New Mexico, Oregon, and Washington all have lava tubes, with the largest share being in California. Most of California's lava tubes lie in the northeastern part of the state. Two hundred and ninety-three individual "caves" have been counted in Lava Beds National Monument, and many lie outside its boundaries. Most of these caves are not individual tubes, but are part of twelve large tube systems.

Tubes form because a stream of lava, like a stream of water, moves faster in the center, where it stays hotter than on the sides. Gradually, the sides cool, forming rock walls. Splashes from the cascading lava add to the wall height, gradually forming an overstream arch. In places where the stream slows, the roof may crust over, pieces of the crust finally joining the sides to form a tunnel with numerous skylights.

Thomas A. Jagger, who studied Hawaiian volcanoes for many years, described the progress of a lava stream through a partly roofed over tube of its own making:

> On the north side of the cone . . . two windows in the roof of the cavern leading from it revealed a brilliantly incandescent cavernous space where, about 12 feet down, the orange-colored melt flowed northward in a majestic torrent. Under the cone, one could see the wellspring which ceaselessly fed the ever-pushing fronts 1½ miles away, and by looking downstream through the window one could see the river pour northwestward in a tube, where blue fume rose through cracks in the roof of hardened flow lava. The entire lining of the tube was at bright-orange heat, and the gases emerging were not particularly sulphurous. This tube had been filled like a sewer when the flow pressure was at its height, and its partial evacuation was evidence of decline in volume of lava.

Because basaltic lava flows are fluid, behaving much like streams of water, lava tubes are commonly sinuous, wandering back and forth across a lava flow like a braided river. In this manner, tubes of two or three stories can be built up, perhaps even cutting across one another.

As a lava stream is slightly arched, lava tubes may form swells in the landscape marking their course. Cave entrances are through natural skylights formed in the discontinuous roof, or through areas where the thin roof has collapsed. On the surface, along the line of the lava tube, spatter cones (hornitos) may mark the former course of the lava stream.

Walking through a tube, one can see that it is oval or round in cross section (although some tubes have slightly different shapes), and that its jointed floor is comfortably flat. Here and there one may come upon a large, bulbous, domed room where lava was held before breaking out again, or one may find stony lakes of congealed lava.

Its walls may show marks of the swash of moving lava. If the tube had been used by a second stream of lava after it had formed, two generations of marks may show. Many tube walls are glazed with natural glass, formed by lava that cooled quickly as the mainstream passed by. In some caves, lava stalactites hang from the ceiling, and cascades of lava may be preserved in stone. Pock marks, formed by gas pockets, as well as splashes from the passing liquid lava, are also preserved in these unusual caves.

In lava country, rivers sometimes drop into lava tubes to run underground for a distance, or they will drop below the ground surface to flow on the top of a buried lava flow that is impervious to water. Such "lost rivers" may reappear farther downstream, apparently bursting from the ground as fountains. Hat Creek, near Lassen Peak, behaves in this manner over parts of its course. When it reaches Burney Falls, water pours over the lip of the falls and from several zones beneath the river bed, forming a curtain of water literally springing from the rock itself. Springs in lava rock at the headwaters of nearby Fall River in Shasta County produce about 900 million gallons daily, making them among the largest springs in the United States.

Most of the tubes in California are in country that is cool in winter but quite warm in summer. For that reason, summer visitors are frequently surprised by the chilly temperatures inside the caves. Some caves, located in wetter country or where the congealed lava blocks allow water to flow through them, are ice caves, which may be hung with icicles or have frozen rivers and frozen waterfalls in them year round. Before electrically powered refrigeration was common, these caves were used as a source of and a storeroom for ice.

Unlike caves in limestone country, lava tubes do not grow slowly through time; they are a one-time creation and should be treated with respect.

Modoc's lava tubes punctuate tablelands—broad, flat-topped plateaus common in volcanic country. Other volcanic tablelands can be seen on the eastern side of the Sierra Nevada, in the Mojave Desert, and in the Sierra itself. One unusual one called "Stanislaus Table Mountain," or sometimes "Tuolumne Table Mountain," rises like a wide wall in the foothills of the Sierra Nevada, dark against the green or golden grass-covered slopes. Its top appears flat, as one sees it from State Highway 108, but rises gently toward the crest of the Sierra Nevada, from which it originated. The dark rock is a form of cooled lava called "latite," a black, fine-grained rock filled with cavities that contained gas when the lava was hot. In some of the cavities tiny crystals of quartz gleam. It is a form of what, in this book, we have called andesite.

This imposing mesa (the Spanish word for table) has been formed in reverse. At one time it was fluid lava, which sought the easiest (and lowest) course down the mountains—in this case, the bed of the Tuolumne River. What billowing clouds of steam must have risen as the lava filled the river, bank to bank! But no person saw it, as it happened over 10 million years ago. So long ago, in fact, that the lava river, which congealed quickly to a tough, resistant rock, now stands above the surrounding land, although in its days of flow it was well below the land surface, ensconced in a stream bed. It has taken those millions of years to erode away the land around it, transfiguring the lava-filled river into a mountaintop.

Huge lava flows like this make horizontal lines dominant in the landscape, but here and there, where cooling of the lava has proceeded at a slow and even pace, prominent vertical lines may develop. The most striking of these are the

FIG. 18. Stages in the creation of Tuolumne (Stanislaus) Table Mountain. *A* shows the Stanislaus River of 20 million years ago flowing through moderately rolling hills, flanked by subtropical vegetation. In *B*, steaming lava has followed the easiest course downhill—the bed of the river. It has filled, but not overflowed, the river valley. *C* shows the ancient river course as it appears today near Knight's Ferry. The lava flow, being much harder than the surrounding hills, still marks the course of the old river. The softer rock of the enclosing hills has been eroded away, leaving a high, sinuous ridge where once was a river valley.

regular columns formed in such fine-grained rocks as basalt. If a large mass of basalt cools slowly and evenly as in a lake or puddle, without the constant churning of downhill movement, it splits into regular patterns along vertical lines called "joints." A six-sided pattern develops when conditions are ideal, but blocks with from three to eight sides can be found. If the mass cools more unevenly, cracks may grow from one another, and are more likely to form at

B

C

right angles. Where the surface on which the lava cooled was level, the columns are straight; where lava had moved over uneven ground, they are bent. Devils Postpile National Monument, near Mammoth Lakes, and Columns of the Giants, alongside State Highway 108, are two scenic examples of such columnar structures (see color photos in center of book).

The postpile at Devils Postpile National Monument has a pattern as nearly regular as any in the world. Here, the percentage of six-sided columns has been measured by two investigators as 45 and 55 percent; compare this with Giants Causeway, Northern Ireland, with 51 percent; Devils Tower, Wyoming, with 32.5 percent; and Craters of the Moon, Idaho, with 16 percent.

At Devils Postpile the regular columns rise like a giant organ out of the San Joaquin River valley. Viewed from the top, the regular pattern of the columns is spread out as if on a parquet floor, now polished by the passage of glaciers.

Scattered throughout southwestern United States are many skeletons of old volcanoes; in some places, the solid volcanic necks remain, the surrounding cones having been entirely removed by erosion. These lend a vertical element to the landscape. Shiprock, New Mexico, and Agathla Peak (El Capitan), Monument Valley, Arizona–Utah, are two prominent ones. In many places in the desert these necks stand alone, isolated from other rock outcrops. Early Spanish residents called them "huerfanos"—"orphans."

Here and there, long ridges run from the tops of the necks or mountains; these were once volcanic feeders, or the sites of flows from the cracked cases of the volcanoes. Shiprock, New Mexico, has a long feeder, now a vertical wall cutting across the landscape.

Most of California's volcanoes are either too young for the terrain to be eroded from around them or so old that the stump of the original volcanic neck is now covered. Some necks, however, are still visible. Morro Rock, near San Luis Obispo, is a 23-million-year-old volcanic neck, one of a chain of fourteen.

Morro Rock is both a volcanic neck and a tombolo (see chap. 8). Landscape features are what they are today, and because they bear the marks of their origin, they are often called by names that tell what they were. This is not to say that landscape features cannot be several things at once. A fault line may be—and probably is—a river valley; a volcano may be a conical hill, a sharp peak, or a coastal headland.

Will California's volcanoes erupt again? Probably. Despite the vast amount of congealed lava visible throughout the state, most of which has erupted during the past 30 million years, the frequency of eruption does not appear to have been greater in these last 30 million years than it has been in the more recent past.

Mounts Shasta, Lassen, or other peaks of the Cascade Range could erupt at any time. When Mount Vesuvius erupted in A.D. 79, it had long been considered inactive; sheep grazed on its grass-covered slopes, and two sizable towns had been built at its foot. The eruption was so unexpected that many residents of the wealthy city of Pompeii lost their lives when hot ash covered them. Pompeii was buried and lost for 1,600 years.

Recognizing the possibility of a similar eruption of active Mount Lassen, the National Park Service, on advice from the U.S. Geological Survey, closed the major tourist facilities at Manzanita Lake in 1975. No eruption was known to be imminent, but an unheralded explosion, such as caused Chaos Jumbles some 300 years ago, could send avalanches hurtling toward Manzanita Lake at speeds of as much as 160 kilometers (100 miles) per hour. The major tourist facilities for the park were located in their potential path. The campground at the lake has since been reopened, but permanent installations, such as the visitor center, museum, lodge, and restaurant, are permanently closed.

In 1979 the Mammoth Lakes-Long Valley area in eastern California began to experience "increased geologic activity." This activity consisted of: (1) recent uplift of the floor of Long Valley caldera; (2) earthquake swarms follow-

TABLE 4. California's Historic Volcanic Activity

When	Where	What Happened
1979–	Mammoth Lakes	Many earthquake tremors, some of volcanic type; may portend volcanic activity
1978	Mt. Shasta	Many small earthquake tremors; may or may not be related to volcano.
1951	Lake City, Modoc County	Mud volcanoes: low cones of mud formed and hot water discharged.
1930	Mt. Lassen	Strong earthquake shocks, presumably originating near Mt. Lassen, were recorded on the seismograph of the Mt. Lassen Volcano Observatory.
1917	Mt. Lassen	Emission of steam: volcano became dormant in 1917.
1915	Mt. Lassen	Great blast: ash and pumice eruptions, new lava formed, mud flows and *nuée ardentes*.
1914	Mt. Lassen	Explosive action: new lava flows and ash falls.
1890	Mono Lake	Sublacustral eruption: emission of steam and sulfurous fumes in puffs; boiling water and hot mud from formerly cold springs.
1857	Mt. Lassen or Mt. Shasta	Ash eruption.
1851–52	Cinder Cone, Lassen County	Surface eruption: development of cinder cone and flows of basalt.
1786	Either Mt. Lassen or Mt. Shasta	Possible steam and ash eruption: explorer La Perouse may have seen eruption while voyaging along the California coast in 1786.

TABLE 4. California's Historic Volcanic Activity (Contd.)

When	Where	What Happened
500 years B.P.*	Burnt lava flow, Siskiyou County	Surface eruption: development of cinder cones and flows of basalt.
500–850 years B.P.	Inyo Crater Lakes, Inyo County	Phreatic explosions: development of crater whose walls are composed of interlayered lava flows and explosive deposits.
1100 years B.P.	Big Glass Mountain, Siskiyou County	Extensive pumice eruptions followed by emplacement of domes: pumice deposits, and flows and domes of obsidian.

*B.P. is the abbreviation for "before the present."

ing four earthquakes of approximately magnitude 6; the epicenters of the swarms—a series of small earthquakes—apparently moved upward from depth, suggesting that magma might be rising; (3) what appeared to be volcanically induced "harmonic tremor" (a continuous release of seismic energy); and (4) increasing vigor of fumaroles. All this activity, coupled with an unusual seismic wave shape, suggested to scientists that subsurface volcanic activity was intensifying in an area less than 5 kilometers (3 miles) from the village of Mammoth Lakes. For this reason, the U.S. Geological Survey issued a "Notice of Potential Volcanic Hazard" on May 25, 1982.

Because this geologic activity was similar to that of Mount St. Helens, Washington, in 1980, and because the type of eruption, if it occurred, could be an explosive rhyolitic one (as has happened in the geologic past), local and state officials made emergency plans in the event of an eruption. By the end of 1983, no eruption had taken place, although tremors were continuing.

Volcanoes that spew out molten rock have provided California with some of its most exciting spectacles, and with elegant features adorning its landscape. But it is molten rock that did not erupt that forms the backbone of the state. Granite and its relatives that were once fluid like lava form the foundation of the great Sierra Nevada, as well as most of the ranges of the southern deserts.

The granite family, including not only true granite but cousins with such mouth-filling names as granodiorite, quartz monzonite, and trondhjemite, are easily distinguished by their pepper-and-salt appearance. Most people recognize granitic rocks because they have seen them so often as stones used in buildings, curbs, memorial stones, and sculpture. The pepper-and-salt aspect comes from the mixture of minerals in the rock: clear, glassy quartz and milky feldspar make the "salt," and bits of black mineral, commonly black mica, make the "pepper." All of these are easily visible because the rock has cooled slowly enough from its molten state to allow the minerals to grow into small crystals. It is not always apparent to the unaided eye that each of these mineral bits is a crystal, because they do not appear to be perfectly formed, as gemstones are. However, a close look with a microscope shows that they are crystalline in nature.

Anyone who has tried to grow crystals knows that the slower they are allowed to grow, the larger the crystals, if other conditions remain the same. This is how granite formed, too: slowly, over a long period of time, with crystals of minerals gradually taking shape in the fluid. Not so for lava: the hot liquid is spilled or thrown out into the colder air so fast that readily visible crystals have no time to form. The two rocks, granite and lava, may have exactly the same chemical composition, but they do not look alike.

Granite, like other crystalline rocks, is hard and tough. Nevertheless, it does weather and erode in time, and through the years has been carved into some of Calfornia's most intoxicating scenery. The artists that carved that scenery are wind, water, and ice, as we shall see in the following chapters.

TABLE 5. Volcanic Landscape Features

Feature	Where to see a good example
Aa	Lava Beds National Monument
	Amboy Crater, Mojave Desert
	Sawmill Creek lava flow, U.S. Highway 395
	Lassen Volcanic National Park
	Lava flow southeast of Baker, Mojave Desert
Ash	See Volcanic ash
Bomb	Sawmill Creek lava flow, U.S. Highway 395
	Cinder Cone, Lassen Volcanic National Park
	Red Hill near Little Lake, U.S. Highway 395
	Mt. Konocti, Clear Lake
	Modoc Lava Beds
	Lava Beds National Monument
	Mt. Shasta
	Dish Hill near Amboy
Columnar lava	Devils Postpile National Monument
	Columns of the Giants, Sonora Pass road
	Interstate Highway 5, near Castle Crags
	Little Lake, Inyo County
Cone (cinder)	Red Cinder Mountain, Owens Valley (cinder cone 600 feet high)
	Red Hill, north of Little Lake, U.S. Highway 395
	Black Point, Mono Lake (basaltic cinders)
	Red Cones, middle fork of San Joaquin River
	Lake Tahoe (cinder cone used for sewage disposal)
	Amboy Crater, Mojave Desert (erupted 6,000 years ago)

TABLE 5. Volcanic Landscape Features (Contd.)

Feature	Where to see a good example
	Cinder Cone, Lassen Volcanic National Park (erupted 1852)
	Schonchin Butte, Lava Beds National Monument
	Deer Mountain, near Mt. Shasta
	Little Deer Mountain, near Mt. Shasta
	Medicine Lake Highland (more than 100)
	Pisgah, 35 miles east of Barstow
	Coso Domes south of Lone Pine, Inyo County
	Kelso, 26 cones on Kel-Baker road southeast of Baker
	Pumice Stone Mountain, near Medicine Lake
	See also map 5
Cone (spatter)	Lava Beds National Monument
Dome	Templeton Mountain
	Silver Peak, Ebbett's Pass (rhyolite dome)
	Markleville Peak, Alpine County (andesite dome)
	Highland Peak, Ebbett's Pass (rhyolite dome with cinder cone on one side)
	Glass Mountain, U.S. Highway 395 (obsidian, 1 million years old)
	Mono Craters (rhyolite tuff rings, dome, and flows, 6,400–10,000 years old)
	Panum Crater (dome with spines, ridges, bombs, and pumice, 1,300 years old)
	Wilson Butte, U.S. Highway 395 north of Deadman Summit (rhyolite)
	Jackson Butte, Golden Gate Hill, McSorley Dome, and Hamby Dome, all near Mokelumne Hill (old, eroded andesite domes)

TABLE 5. Volcanic Landscape Features (Contd.)

Feature	Where to see a good example
	Lassen Peak, Lassen Volcanic National Park (large plug dome)
	Sutter Buttes, near Marysville
	Black Butte, near Mt. Shasta
	Salton Domes, Salton Sea
	Pumice Buttes, Salton Sea
	See also Obsidian flows
Explosion pits	Inyo Craters, near Mammoth Mountain, U.S. Highway 395 (500–800 years old)
	Paoha Island, Mono Lake
	Ubehebe Craters, Death Valley National Monument
	Devil's Punchbowl, Mono County, near U.S. Highway 395
Lahar (Volcanic mud flow)	Two Teats, Mt. Morrison 15-minute topographic quadrangle (3 million years old)
	Carson Spur, State Highway 88
	Thimble Peak, State Highway 88
	Lassen Volcanic National Park
	Mt. Shasta
Lava flow	Devils Postpile National Monument (columnar structure; parts of flow show pillow structure where lava flowed into water)
	Oroville Table Mountain, Butte County (basalt) Tuolumne
	Tuolumne (Stanislaus) Table Mountain (good view at intersection of State Highways 108 and 120 (9 million years old)
	Sonora Peak
	Leavitt Peak
	Sawmill Creek, U.S. Highway 395

TABLE 5. Volcanic Landscape Features (Contd.)

Feature	Where to see a good example
	San Joaquin River headwaters, especially near Pincushion Peak and Saddle Mountain (2–4 million years old)
	Mt. McGee, Deadman Pass (2–4 million years old)
	San Joaquin Mountain, John Muir Trail
	Dardanelles, Sonora Pass, State Highway 108
	Columns of the Giants, Sonora Pass, State Highway 108 (150,000 years old)
	Glass Creek obsidian flow (rhyolite glass)
	Lava Beds National Monument
	Lake Britton (north edge)
	Warner Mountains
	Medicine Lake Highland
	Glass Mountain, Siskiyou County (obsidian flows, some less than 1,000 years old)
	"Earthquake fault" near Mammoth Lakes
	Many flows in southern deserts
Lava tubes ("caves")	Lava Beds National Monument (more than 500 tubes)
	Subway cave (north of Lassen Peak on State Highway 44)
	Pluto's Cave, near Mt. Shasta
Maar	Inyo Craters, Sierra Nevada
Mud pots	Lassen Volcanic National Park
	Salton Dome, Salton Sea
Obsidian flows	Medicine Lake Highland, esp. Glass Mountain, Siskiyou County
	Little Lake, Inyo County
	Glass Creek, Mono County
	Clear Lake, Lake County

TABLE 5. Volcanic Landscape Features (Contd.)

Feature	Where to see a good example
	Salton Sea, south end
	Mammoth Lakes, Mono County
	See also Domes
Pahoehoe	Modoc Point
	Amboy Crater and Pisgah Crater, Mojave Desert
Pumice fields	Pumice Stone mountain, Siskiyou County
	Devils Punchbowl, Mono County
	Clearlake Highlands, Lake County
	Sotcher Lake and Pumice Flat, Devils Postpile National Monument
	Pumice Buttes, Salton Sea
Volcanic ash and tuff (remnants of nuée ardente)	Buena Vista Peak and Valley Springs Peak near Valley Springs, Amador County (20–30 million years old)
	Quarry east of Altaville on Murphys Road (20–30 million years old)
	Exposures on U.S. Highway 395 near Bishop, along shores of Lake Crowley and Owens and Rock creeks (Petroglyph Loop Trip petroglyphs are carved in 700,000-year-old tuff)
	Sotcher Lake, Devils Postpile National Monument
	Reds Meadow Ranger Station, Devils Postpile National Monument (welded tuff; very hard—sealed together as it cooled)
	Mt. Konocti, Lake County
	Surprise siding, Modoc County
	Mt. Shasta
	Pinnacles, near Alturas
	Modoc Point

TABLE 5. Volcanic Landscape Features (Contd.)

Feature	Where to see a good example
Volcanic neck (eroded)	Morro Rock, San Luis Obispo County (23 million years old; westernmost of 14 necks)
	Moapi, central Turtle Mountains, Mojave Desert
	Lover's Leap, Pacheco Pass, Santa Clara County
	Eagle Rock, near Tahoe City
	Pinnacles National Monument
	Club Peak, east of Baker, Mojave Desert
Volcanic tablelands	Stanislaus (Tuolumne) Table Mountain (inverted)
	Bishop tableland (can be seen well on Petroglyph Loop Trip, along U.S. Highway 395)
	Oroville Table Mountain
Volcano	Mt. Rose, Nevada
	Mammoth Mountain, near U.S. Highway 395
	Mt. Shasta (stratovolcano)
	Mt. Lassen (stratovolcano)
	Mt. Konocti, near Clear Lake
	Sutter Buttes
	Glass Mountain
	Little Glass Mountain
	Negit, Mono Lake
	Mt. Harkness, Lassen Volcanic National Park (shield volcano)
	Red Mountain
	Prospect Peak, Lassen Volcanic National Park (shield volcano)
	See map 5

5 • WATER—AGENT OF CHANGE

Water has been the principal sculptor of California's landscape. Working with rocks derived from volcanoes and from deep within the Earth, or with rocks formed on or near the surface, water carves and alters the shape of the land, etching deep canyons here, building beaches there, sanding this rock into a turret, that into a pyramid. While water is altering and moving rocks of all sizes, soil is being formed and new rock layers are beginning. These processes are all at work at once.

Water works in all its forms to transfigure the landscape—as vapor, as liquid, as ice. Water's ability to accommodate itself to the shape of the place in which it finds itself is one of its unusual—and life-sustaining—properties. Because it can take the shape of pores within the rocks of Earth or be heaped into clouds in the sky; because it can fall as delicate snowflakes to be metamorphosed into grinding glacial ice; because it can move swiftly from place to place under no force save that of gravity; because it is as malleable as the tides and yet virtually indestructible—for all these reasons, water has become the distinguishing feature of our planet.

Fortunately for us, water is constantly moving, changing its form. This perpetual movement has been called "the hydrologic cycle," and upon its continuance our lives depend.

The major storage area for the water that gives our "blue planet" its name is the ocean, which wraps about the Earth,

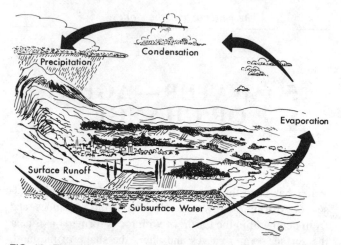

FIG. 19. The water cycle is the pattern of water movement as it circulates through Earth's natural system: precipitating from the atmosphere to the Earth as rain and snow, running on the surface as streams flowing to the sea, percolating into the subsurface and seeping back up to the surface, and finally being evaporated by the sun and transpired by plants back to the clouds, to begin the cycle again.

interrupted here and there by continents. If all the water in the seas were somehow stacked on the surface of California, it would be 150 kilometers (93 miles) deep. From this vast reservoir of the oceans, the sun, which powers the hydrologic cycle, draws water toward it by evaporation to form clouds. From the clouds, rain and snow return to the Earth.

Of the 425,000 cubic kilometers (102,000 cubic miles) of water that is evaporated annually, 325,000 cubic kilometers (78,000 cubic miles) falls back into the ocean. The remainder falls on land to serve as the freshwater supply for animals and plants. About 38,000 cubic kilometers (9,000 cubic miles) of the water that falls on land is returned to the sea by rivers and streams. The rest is either used by living creatures or soaks into the ground to be stored for later use. It, too, is eventually returned to the atmosphere. Some of the water in the ground runs by gravity toward the sea, but most flows toward streams where it joins the surface flow that eventually reaches the sea.

Thus, water is always moving. It may stay in the ground for thousands of years, or may be halted in lakes or glaciers; but sooner or later, it is released to complete the cycle and begin again.

Seas of the past have contributed sediment toward construction of the land, and seas of today wear away its edges; but, except for acting as a reservoir for water in the hydrologic cycle, today's seas have little effect on the present features of the land except along the thin strips of coastline. Nor does ground water, by its very nature, reach the surface to change it. Glaciers, too, have had their day in California's landscape; those that remain are making but small changes.

Yet water is the principal sculptor today, as it has been in ages past—not as oceans or glaciers or ground water, but as streams, rivers, lakes, and thin films of moisture.

The thin films of moisture prepare the way for more vigorous action by water. They penetrate solid rock to groom the rock fragments for their roles both as tools of erosion and material transported by the stream. The fragments and the parts of the rock dissolved by water ultimately become new layers of rock or new soil; they commence their transformation and are brought to their new resting places by water.

TABLE 6. California's Rain and Snow

In an Average Year
- 200 million acre-feet falls, which is equal to 65 million million gallons

- 65 percent of this is lost by transpiration and evaporation

- 70 million acre-feet, or 23 million million gallons, is available for human use

- Of this, 40 million acre-feet, or 13 million million gallons, is actually used by agriculture, industry, and domiciles, and 30 million acre-feet, or 10 million million gallons, goes to the sea

Sometimes enough water enters cracks or pores in the rock to turn to ice when cold weather comes; pressure from the freezing ice may dislodge particles of rock which later are carried by gravity or rainwash to the streambed to form part of the stream's "load" (the material a stream carries).

Or, if the climate is not cold or the film too thin, water may linger as a liquid to start a chemical reaction within the rock. Any mineral may be attacked, but feldspar minerals are among the first to succumb. Because feldspars are so widespread, their degeneration is the single most important item in the weathering process. What happens during weathering—the process of chemically breaking down rock—is:

feldspar + water = clay + solution

The resulting clay (of which there are many types) and solution is larger in volume than the original feldspar plus water. For that reason, minerals adjacent to feldspar crystals are broken apart in the process of clay formation. When clay minerals come in contact with water (as in this reaction), they swell to several times their dry volume. One type of clay, called montmorillonite, will swell to as much as fourteen times its size when dry. When it swells, the rock itself breaks apart, as does any building that may be built on it!

All of this broken material moves downhill, generally toward the nearest streambed. It may take centuries or it may be flushed down by a storm in minutes, but ultimately the loosened material finds its way to a stream. Once there, the rock fragments continue downhill by the force of water. Unless halted at a low spot en route, the fragments, like the stream, seek the sea. Large rocks, some as large as houses, are bounced or rolled along the bottom; sand-sized particles move as a curtain near the bed; very fine particles, as well as uncommonly flat ones, are carried within the body of the

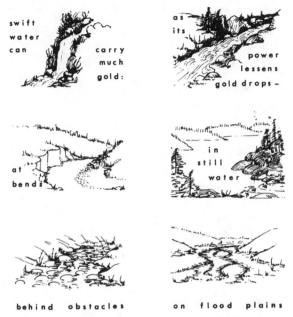

swift
water
can   carry
                much
                gold:

as
its
            power
         lessens
      gold drops –

at
bends

in
still
water

behind   obstacles   on   flood   plains

FIG. 20. How a stream transports its load. The example here shows gold being carried, but a stream carries all of its visible load in this manner.

water in suspension; and, of course, some particles have been dissolved.

To some extent, water uses this rock material as sandpaper to wear away other rocks on the bottom and sides of the stream; and to some extent, the rock fragments themselves are broken into smaller pieces by tumbling over and over in the streambed.

It is surprising how much rock material water can transport. In spite of the fact that 95 to 97 percent of the energy of a river is lost in turbulence and friction, the remainder of its power is enough to move mountains. The Mississippi and the Colorado rivers are the two leading carriers per unit of drainage area in North America, the Mississippi carrying

MAP 6. California rivers.

an average of 477 tons of sediment per square mile per year, and the Colorado 438 tons. This counts only the dissolved material (23 percent for the Mississippi, 12 percent for the Colorado) and the suspended matter. It does not take into account the boulders and pebbles that are bounced along the stream bottom.

Since the loads these rivers carry originate in the moun-

tains, the rivers are actually wearing the mountains down—
making the "rough places plain." It has been estimated that
the Sierra Nevada is now being worn down by its rivers at
the rate of an inch per thousand years, while the more active
Eel River is carrying away the Coast Ranges at the rate of
100–200 mm (40 to 80 inches) per thousand years. This is
15 times as rapidly as the Mississippi is eroding its basin.

Rivers flow quietly or wildly, over stills and rapids, into
pools and over bars on their downhill course. How they flow
changes from day to day with the weather, and from place to
place as the streambed varies. Water that moves slowly
along a smooth, straight channel moves by laminar flow, in
which the upper layers move somewhat faster than the lower
ones, with a maximum speed slightly below the water sur-

FIG. 21. A rushing stream.

face. This type of flow is rare in streams; it is so slow that virtually nothing is transported.

More common is turbulent flow, in which water seems to tumble over and over itself. Turbulent flow can be either streaming or shooting (as in rapids). These two types of flow accomplish most of a river's work.

How much material—sand, gravel, boulders, and tiny particles in suspension—a stream can move depends upon the velocity of the stream, the amount of water in it and the roughness of its bed, as well as the type of fragments it is attempting to carry.

A stream gathers these particles in several ways: some are washed into it (such as weathered rock fragments loosened by frost and rain) and some it derives by eroding its banks. All the particles it picks up start their long journey toward the sea. Some particles may make a quick trip; others may take thousands of years; still others may be waylaid en route so long that the river itself vanishes.

California has a number of vanished rivers. In the Sierra Nevada, streams that once flowed through the savannah of 30 million years ago can still be traced, showing the pattern of drainage in that far-off time. The river beds were preserved intact by being filled with a layer of ash. The gravels of these ancient river bottoms are still to be seen in the Sierran foothills—or were. In many places, the old streams have been disinterred, and the gravel robbed of its gold by California argonauts.

Rivers themselves take definite patterns. If the rock over which they flow is of uniform hardness and if the slope of the land is generally uniform, the pattern made by the river and its tributaries assumes a shape like that made by the branches of a tree. But if, for example, there are layers of harder rock mixed with more easily eroded ones, a stream may be deflected by the harder rock, changing its pattern. This overall pattern, which one can see only from the air or on a map, is the pattern of the whole drainage network as a unit. The size of a drainage basin depends upon one's point of view. For example, the drainage basin of the Sacramento–San Joaquin River system includes most of the Sierra Nevada and part of the northern volcanic region; the Sacramento River alone drains the Sacramento Valley, and includes tributaries from the northern Sierra and the volcanic region; the Feather River, one of the tributaries of the Sacramento, drains a smaller part of the Sierra Nevada. It, in turn, has tributaries, each of which has a drainage area of its own.

Each river, too, is a landscape form in itself. Viewed from above (say, from an orbiting satellite), a river's path appears erratic, with a few short, straight stretches, zones of meander, and possibly "braided" areas (areas of crisscrossing channels) near its mouth. Most rivers look as if they have more space than they need for the amount of water that flows in them; they flow only in a portion of the area that appears to be claimed by them. What we are seeing is the floodplain of the river—the total area it covers when the river floods its channel.

If we were to look at the cross section of a river we would note that the area it occupies has a flat "D" shape. The deepest part (called the "thalweg") is sometimes on one side, sometimes on the other (giving the D a rakish aspect), but occasionally in the center. If the river has greater flow, the size of the D increases, but its shape remains the same.

Looked at in side view, the long profile of a river may resemble the profile of a saucer, going generally downhill, but with a long flat area near its end. The shape is far from even, as anyone knows who has canoed a river from beginning to end. (Curiously, the beginning of a river is its "head"; the end is its "mouth.")

The general downhill course is interrupted by waterfalls and rapids at the upper end and by pools and riffles where the water is quieter. Pools (quiet spots) and riffles (rough spots) alternate at a roughly even spacing where the river is carrying fine, uniformly sized material. Where its bed load is a mixture of coarse and fine material, the alternation of pools and riffles is less rhythmic. At high water, pools and riffles are no longer visible; they have been "drowned out" so that the water appears flat.

Here and there water forms eddies, tiny whirlpools drawing circles in the water. Rock and sand caught in an eddy can gradually grind into the rocky bed of the stream itself, creating a stream-made mortar, or "pothole."

The river's third dimension—its shape when viewed from above—is easiest to visualize, as we are used to seeing it on maps. Viewed this way, rivers are seen to have some areas that are fairly straight, others that are remarkably sinuous, and others that seem tangled. The symmetrical, sinuous portions are "meanders," the more tangled portions, "braids."

Why meanders form is not altogether understood, but they do form in less steep stretches than braids; they are most sinuous where they run over a uniform surface and where the river is carrying a uniformly sized load. One often reads that meanders begin when rivers encounter a slight obstruction, which causes the water to swerve; this, in turn, is compensated by a swing back, and so on. However, this does not explain why meanders are so beautifully regular, or why meanders form on the most even of surfaces. You can prove this by dropping water on a slightly tilted sheet of glass; if the tilt is not too great, meanders will form. There are even meanders in the Gulf Stream—meanders of water within water.

Laboratory experiments show that when either the gradient (slope) of the bed or the amount of water in the stream is increased, the meanders become wider (that is, their radii become greater). As one might expect, streams winding through finer, more uniformly sized material have more uniform bends. Fine, resistant material encourages a deeper channel with a gentler gradient; coarser material allows wider, shallower channels with steeper gradients. The river—in this case a laboratory river—is obviously adjust-

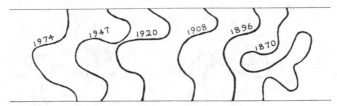

FIG. 22. Meanders of the Sacramento River through a century. The river has been generally straightening its course, and shortening its length.

ing to its surroundings. A stream that meanders uses less energy than one that does not; therefore, the easiest course may be a winding one.

Whatever causes meanders, they are meaningful features for a river. If you look at a map without a scale, it is difficult to tell large rivers from small ones; the pattern of sinuosity is the same.

One good place to observe symmetrical meanders in California is on a tidal mud flat. Here, where the level of the ground is sloping slightly, and the surface is fine grained and uniform, tiny streams show meanders of beautiful regularity. They are visible at both high and low tide, but must be viewed from above at low tide, as the cover of water "drowns" them.

Mud flats are also a good place to observe braiding in miniature. Here, one can see the features of a braided river: water courses that converge and unite (hence the word, "braided"). Since mud flats include river floodplains, as well as those that are tidally flooded, it is possible to see most floodplain features here: meanders, braided river courses, meander scrolls or oxbows (portions of a meander that are symmetrically curved like the wooden cross-piece formerly used to hitch oxen), oxbow lakes (lakes that were once part of a meander but have since been shut off and abandoned), sloughs, natural levees, and swamps.

These same features may be seen in the floodplains and at the mouths of large rivers, of which the Sacramento is a

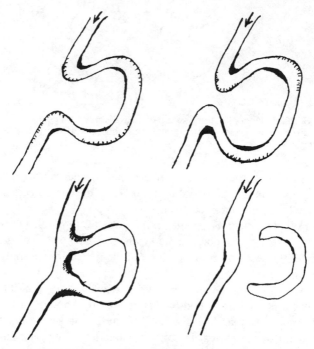

FIG. 23. How cutoffs and oxbow lakes are formed.

classic example. For boating and fishing enthusiasts, the Sacramento–San Joaquin rivers provide more than 1,600 kilometers (a thousand miles) of braided waterways to explore. However, when one is actually on the river, it is not possible to appreciate the braided effect; it is only from the air or on a map that one gains sufficient perspective.

Braided channels are wide, and form in those reaches of the stream that have greater gradient and a greater amount of water than meandering portions. Because the amount of water is constantly changing, depending upon conditions upstream, and because the river itself can adjust its gradient, the river's pattern also can change through time.

It is the ability to change rapidly that sets rivers apart from other landscape features and landscape-changing

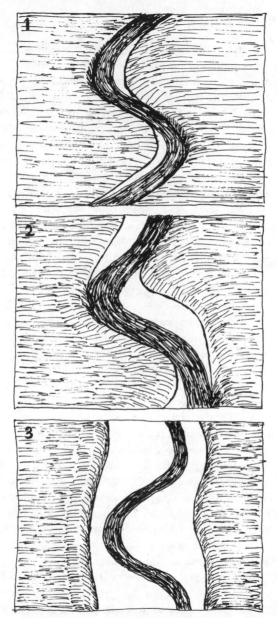

FIG. 24. How a river widens its floodplain.

agents. Earthquakes along faults lift mountains suddenly, but it takes many earthquakes and many thousands of years to raise a range as high as the Sierra Nevada. The sea beats on land, altering it mightily in great storms, but making only slow but relentless progress the rest of the time; glaciers make vast changes, but in a ponderously slow manner; the wind makes small changes quickly and often, but its total effect is slight. Rivers, however, wear the land as relentlessly as the sea and as determinedly as glaciers, but through mercurial changes that are their own hallmark.

Most Earth features are in the process of attaining equilibrium. An overhanging cliff may adjust in an instant by crashing down when slight additional strain is added; in contrast, massive rocks may require hundreds of years to adjust to reduced pressure when glacial ice has melted from above them. Streams, however, can adjust speedily to changes in flow, changes in slope, changes in load, and changes in the arrangement and type of rock over which they flow. Rivers in perfect balance are "graded"; that is, they choose paths that will give the precise gradient needed to support their loads. Meanders lengthen a stream's course and lower its gradients (slope); for this reason, a meandering stream uses less energy.

Rivers do not attain perfect balance immediately. There are times and places where the stream's load is too great for it to carry. Perhaps flood waters have subsided and the stream no longer has the flow necessary to transport the load of debris it has assumed; it will then drop as much as necessary for the new conditions. The dropped material remains there until conditions change again, when perhaps the river can once again move the load onward. Or, instead, a new load may be dropped on top of the abandoned one, thus building up a succession of layers.

A glance at map 6 will show that the San Joaquin River lies much closer to the Coast Ranges than to the Sierra Nevada. It has been forced there by its tributaries, which take their water from the High Sierra, rather than the Coast

Ranges. Each of these streams has, since the time of the Great Ice Age, carried a large load of sediment worn from the mountains by glaciers and by streams. When each tributary reaches the flat valley, it can no longer carry as heavy a load, and must—and has—dropped much of it to form a large, alluvial apron along the west flank of the Sierra Nevada, forcing the San Joaquin River to flow closer to the Coast Ranges. During much of the second half of the last century steamboats plied the San Joaquin River from Stockton to Firebaugh or Herndon and the Merced to Cox's Ferry. They carried heavy loads of groceries, lumber, posts, and other supplies upriver and hides, tallow, and tens of thousands of sacks of grain downriver. Passengers were picked up at any point along the stream. Navigation was easy in fall, winter, and spring but became impossible after the early summer snow melt until fall rains again filled the streams.

In this century, most of the tributary streams have been dammed, so that they no longer carry a heavy load into the valley. Nor do they bring much water; there is little left of the San Joaquin to flow to the sea.

The Sacramento River, in contrast, flows closer to the Sierra Nevada. Its tributaries have not forced it out as far. They take their origin from lower mountains, for the Sierra Nevada becomes progressively lower as one moves northward.

At times, the upstream portions of a river may be steepened, thus increasing the slope. If the river is so settled in its bed that it cannot form new meanders to lessen its gradient, it can respond to this new situation by scouring—eroding part of its bed so as to carry a heavier load. If the river is able to etch its bed more deeply, it may continue doing so until it forms a deep canyon such as Kings Canyon in the Sierra Nevada or the Grand Canyon of the Colorado. In some places, streams have settled themselves into their niches, leaving their old floodplains as "terraces" above the new level. River terraces that mark a change in the river's

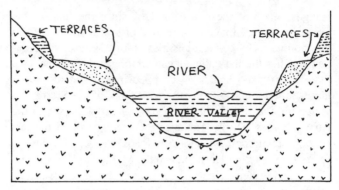

FIG. 25. Terraces of a river valley. Highest one is oldest, the lowest youngest.

condition can be seen along the Klamath River of northern California.

Sometimes one river is able to erode its bed much more rapidly than an adjacent one. Its extra strength may be caused by greater amounts of water or by renewed vigor; whatever the cause, the stronger stream may "steal" the waters of the other river, leaving it as a dried ghost of its former self. Geographers call this act "stream piracy"; where the empty riverbed cuts through the mountain a "wind gap" is left.

Where rivers enter the sea, the shore can be greatly changed. This is where bays and estuaries form, where deltas are built, and where sand and mud are carried down to build the beaches. Rivers commonly form deltas on reaching a body of still water. Here, the river loses its identity, merging with the quieter water; consequently, all the sand and mud it has been carrying are dumped at the river mouth, forming a prism of debris shaped like the Greek letter "D" or delta (Δ). Upstream, the river leaves deposits in its floodplain, as well as bars in the stream itself.

California's best-known delta lies not at the sea where most large deltas are, but inland, where the Sacramento and

San Joaquin rivers merge to enter the chain of waters that eventually leads to San Francisco Bay and the sea.

The load that rivers carry and the erosive work that they do is greatly increased in California by landsliding. Landslides and their relatives—rock falls, debris slides, mudslides—are triggered by an increase in stress. Generally, that stress is derived from an increase in water (which is why there are more slides in the rainy season than in dry weather), but it can also be triggered by earthquakes and other sudden changes.

A frequent cause of landsliding in California is the undercutting of slopes for construction. Slides that take place for this reason are more expensive than most other slides

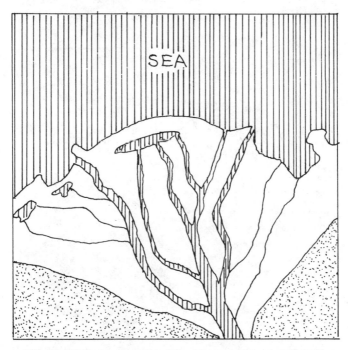

FIG. 26. Bird's eye view of a delta, where a river enters the sea.

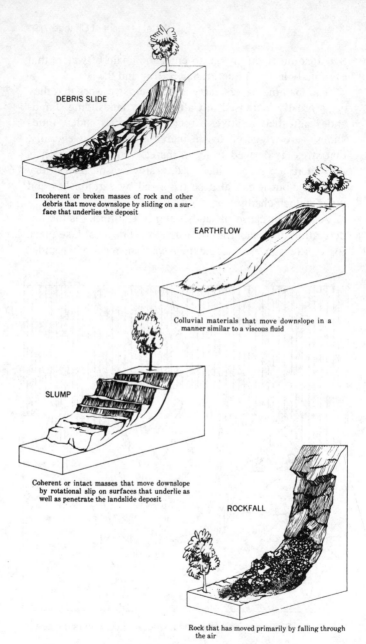

DEBRIS SLIDE

Incoherent or broken masses of rock and other debris that move downslope by sliding on a surface that underlies the deposit

EARTHFLOW

Colluvial materials that move downslope in a manner similar to a viscous fluid

SLUMP

Coherent or intact masses that move downslope by rotational slip on surfaces that underlie as well as penetrate the landslide deposit

ROCKFALL

Rock that has moved primarily by falling through the air

FIG. 27. Types of landslides.

because there is usually a road or building beneath them to be damaged or destroyed.

Los Angeles County, plagued by landslides, instituted stringent regulations for construction in landslide-prone areas. The result was dramatic: millions of dollars reduction in losses from landslides after enforcing the new regulations.

Although many landslides are triggered by human activities, landslides are a natural phenomenon, as anyone who drives through California's north coastal area can see. Here, in addition to the several slides marked by road signs, those who are familiar with the rumpled look of landslide country (reminiscent of Salvador Dali's "soft-watch" paintings) will note that much of the north coastal area is covered by old and new slides.

Some slides move slowly and relentlessly, a few inches or less a year; others burst loose at speeds in excess of 200 kilometers (120 miles) per hour, traveling on a cushion of air. Such a landslide took place in prehistoric time in Blackhawk Canyon, San Bernardino County, California. In 1970, an avalanche of this type in Peru was triggered by a 7.8 magnitude earthquake. The slide buried two towns, killing more than 20,000 people.

During the winter of 1977-78, heavy rains in southern California caused many slides of the type called "debris flows," some of which overtopped the dams created to hold them. The town of Hidden Springs was wiped out, several people were killed, and the area suffered millions of dollars worth of damage. In one grisly episode, thirty bodies were torn from graves in the Verdugo Hills Cemetery and washed into the streets and yards of Tujunga.

In 1956, expensive hillside homes in the Portuguese Bend area of the Palos Verdes Hills, Los Angeles County, began to slide slowly downhill, breaking apart as they moved. Although the landslide had been mapped by geologists years before, many homes had been built upon it. By 1968 damage had reached $10 million; the sliding still continues.

Another continuously moving landslide may be seen at Wrightwood, a resort area in the San Gabriel Mountains. Here, a moving mountain mass of some 18 million cubic yards threatens the town.

One landslide in northern California has caused a part of State Highway 1 to be closed permanently; another within the city of San Francisco prevents use of the scenic highway near Land's End.

A study of landslides in nine San Francisco Bay area counties has pinpointed the location of hundreds of slides. In the wet winter season of 1968–69, landslide losses in these counties was tagged at more than $25 million. Unless vigorous action is taken, a State of California report says, the state can expect to have total monetary losses of almost $10 billion as a result of landsliding from 1970 to the year 2000.

The January 1982 storm that poured more than a foot of rain on Mount Tamalpais in 32 hours started hundreds of landslides in the already rain-soaked hills. More than 80 houses were destroyed by slides in Marin County alone; in Santa Cruz County, a mud slide on Love Creek buried many houses with people in them. At least thirty-one persons died, and five were missing, presumed dead. Financial losses were estimated in the first few days to be $161.1 million worth of private property plus $89.9 million in public property, for a total of $251 million from just this one storm.

Given time enough, the material of which slides are made usually finds its way to rivers and streams to be washed toward the sea. Two exceptions to this rule are slides in the desert, which may not move toward the sea until the climate changes, and slides on the coast which drop directly into the sea.

All of the things that water does to reduce the landscape—weathering rocks, transporting sand and gravel, wearing down river banks, encouraging slides, depositing debris—all of these wear down the mountains, tending to reduce the land to a level plain. Whether or not a level plain is ever produced depends upon how vigorous the action of water is (which, in turn, depends upon climate), how hard and resistant the rocks are, and whether or not the water can reduce the land as fast as other forces build it up.

While the land is being changed by water from one shape to another, water itself provides some of the most interesting landforms, including rivers and streams, waterfalls, lakes, ponds, and marshes.

Waterfalls, as everyone knows, form in steep parts of the landscape; they cannot form where the land is even or flat. Although waterfalls are formed for a host of reasons, in California only a few types of waterfalls are represented.

Most spectacular are the falls of glacial origin, including all the falls of Yosemite and Hetch-Hetchy valleys. Here, the passage of thick glaciers has cleared deep canyons over whose lips water plunges in astonishing displays. In high country, glaciers have cut "staircases" now connected by waterfalls; and at what was once the glacier's head, waterfalls plunge into deep bowl-shaped depressions called cirques.

Another type of waterfall in California is seldom more than 30 meters (10 feet) high. Less spectacular than glacial ones, this type was caused by water plunging over a large joint block of granitic rock. Waterfalls similar to this in appearance may form wherever there is a hard mass of rock in the stream's path, but since so much of the high Sierra is granitic, they are the most common type. If a stream has not yet settled its course, it can steer around hard masses of rock; but after it has eroded a valley for itself, it can worry its way deep enough in its bed to encounter a hard lump of rock, but can no longer seek another path. Such a hard rock, being less easily eroded than the remainder of the bed, becomes an obstruction over which the water must leap.

California has few of the falls common farther east in the United States: waterfalls that develop when a stream wears its way down to a hard, nearly horizontal stratum of rock—

as is dramatically apparent at Niagara Falls. In California, few rock strata are horizontal, thanks to mountain-making movements that have been more or less continuous these past millions of years.

There are, however, young lava flows that are essentially horizontal. In places, these have served as hard ledges for waterfalls. An unusual falls of this sort is Burney Falls, where water has percolated into the lava and runs underground, finally spewing out from within the face of the waterfall as well as from its lip. The force of water here has created a small lake or plunge pool at the base of a waterfall, as is common in other falls. Such pools are among the host of large and small lakes with which California is blessed.

Over a thousand of California's lakes are man-made reservoirs that store more than 39 million acre-feet or nearly 13 trillion gallons of water. The remaining lakes in California are natural, and owe their origin to an assortment of geologic causes. The great majority of the natural lakes are glacial lakes, most of them small tarns (see chapter 6).

All of the state's large natural lake basins have been created by mountain-making movements. Largest of all, the Salton Sea, lies between two fault blocks. Clear Lake, the largest freshwater lake wholly within the state, lies in a basin that was warped downward and blocked by the action of volcanoes; Lake Tahoe lies in a down-dropped block dammed by volcanic flows. Mono Lake, a relic of the wetter days of the Great Ice Age, occupies a nearly circular depression created by earth movements. In addition, a host of smaller ponds and puddles called "sag ponds" stand in the rifts of the state's major fault zones.

Some lakes have been created by landslide dams: such a lake is tiny Mirror Lake in Yosemite Valley, dammed by a rock fall.

Many of California's "lakes" are lakes no longer, but are playas, dry most of the time (see chapter 9). Some are

MAP 7. Lakes and reservoirs of California.

merely vernal pools that have water in the spring, but dry
when the wet days have gone. In addition to these, there are
many ephemeral lakes that hold water sometimes, but are
frequently dry. Goose Lake, source of the Pit River in
northern California, is such a lake. Wet or dry, Goose Lake
looks like a lake; in contrast, it is very difficult to see the

huge ephemeral lake, Buena Vista, if it is dry, as growing crops cover it. Before dams were built on its headwaters, it was vast and shallow in wet years and still is, in flood.

Most plentiful of California's lakes are those created by glaciers. Some of them have been dammed by moraines left from ancient glaciers; the basins of others have been scoured out by the ice itself on its downward path. Some have been left on highlands, some in glacial trough floors, and some in rock stairways as parts of chains of lakes.

Volcanoes have been the cause of some lakes. Hat Lake in Lassen Volcanic National Park was created when a volcanic mudflow built a dam; Medicine Lake in the Medicine Lake Highland was formed in a similar way. Snag and Butte lakes in Lassen Park trace their origin to a lava flow that provided a natural dam.

Rivers, also, have created lakes. A few river-made oxbow lakes may be seen in California, but on the whole, the state is too mountainous to have many features of this sort.

However, one of California's largest lakes, the Salton Sea, is in part river created. A few thousand years ago, a large freshwater body, Lake Cahuilla, occupied much of the Salton Trough, a basin formed by faulting. Lake Cahuilla had formed when the Colorado River built its delta at the north end of the Gulf of California high enough to form a dam. Behind the dam Lake Cahuilla rose, but slowly dried up as the climate grew warmer and drier.

By the time of the forty-niners, the basin of former Lake Cahuilla, which was many feet below sea level, held the reputation as the hottest desert the immigrants must cross. It was so hot that few of them chose the route leading through it. Then, in 1904–1907, the Colorado River overflowed its banks to form a new lake—the Salton Sea. It, too, would dry in the hot desert air were it not for waste irrigation water that finds its way into it.

Water running over the land or ice moving on it have carved many of California's landscape features. A few, however, have been etched out in the dark underground: the limestone caves of California. The caves have been formed by ground water, which, as it penetrates the soil, adds carbon dioxide (CO_2) to its composition, thereby becoming weak carbonic acid. The acid dissolves limestone, carrying it in solution along underground passages, enlarging them. Eventually, a large cave system may be formed underground, as it was in the midcontinent region of the United States.

California limestone caves are small, compared to the giants of the Midwest, and most of them are far from level. The carbonate rock in which they have been etched was formed in seas of Paleozoic and Mesozoic times, generally more than 150 million years ago. Although the layers were level when the rock was first formed, California's mountains are almost all younger than its limestone. The limestone beds have been twisted, crumpled, broken, and eroded as the mountains were built, so that they are oriented in all directions, and have taken a variety of shapes. Caves etched into them reflect these strenuous times.

Thus water is both landform and land former. Water in rivers and lakes, water in seas and ice, water in falls and pools provides our most treasured landforms; yet that same water, as well as water underground and water in the air, is the transformer of land, reshaping everything it comes in contact with—even its own bed.

TABLE 7. California Lakes

Origin of Lake	Where to see a good example
Dry lake (playa)	See map 18
Ephemeral (intermittent) lake	Buena Vista, San Joaquin Valley
	Tulare, San Joaquin Valley
	Goose, Modoc County
	Elsinore, Riverside County
Fault-block lake	Tule Lake
	Lower Klamath Lake
	Marlette Lake (Sierra Nevada)
	Honey Lake (playa)
	Alkali Lake (playa)
	Goose Lake
	Mono Lake
	Owens Lake (playa)
	Searles Lake (playa)
	Panamint Lake (playa)
	Death Valley playa
	Salton Sea
	Saline Valley playa
	Lake Elsinore
	Lake Tahoe
	Deep Springs Valley Lake
	Soda Lake (Carrizo Plain)
Glacial lake	
Ice-scoured lake on a glacial trough floor	Silver Lake (American River)
	Loon Lake
	Helen, Emerald, Shadow, and Cliff lakes, Lassen Volcanic National Park

TABLE 7. California lakes (Contd.)

Origin of lake	Where to see a good example
Lake formed in the end of a glacial trough (dammed by moraines)	Donner Lake
	Reversed Creek Lakes
	Twin Lakes (Mono County)
	Walker Lake
	Fallen Leaf Lake
	Convict Lake
Lake on ice-scoured highlands	Lakes in Humphrey Basin, Sierra Nevada
	Desolation Lake, Sierra Nevada
	Tulainyo Lake, near Mt. Whitney
Step lake	Little Yosemite Valley
	Black Rock Pass
Glacial cirque lake (tarn)	Mary Lake
	Garnet Lake
	1000-Island Lake
	Gold Lake
	Many other lakes in the high Sierra Nevada
	Dollar Lake (San Bernardino Mts.)
	Gray Lake (San Bernardino Mts.)
Hot spring lake	Boiling Springs Lake, Lassen Volcanic National Park (in a glacial basin)
Landslide lake	Blue Lake
	Kern Lake
	Manzanita Lake, Lassen Volcanic National Park
	Mirror Lake, Yosemite National Park
	Zaca Lake, Santa Barbara County

TABLE 7. California Lakes (Contd.)

Origin of lake	Where to see a good example
Playa	See map 18
Ox-bow lake	Pear Slough, San Joaquin River
	Murphy Lake, Feather River
	Horseshoe Lake, Sacramento River
Volcanic lakes	
Dammed by volcanic flow	Butte Lake, Lassen Volcanic National Park
	Snag Lake, Lassen Volcanic National Park
Dammed by volcanic mud flow	Hat Lake, Lassen Volcanic National Park
	Medicine Lake, Medicine Lake Highland
	Thurston Lake, Medicine Lake Highland

TABLE 8. Features Formed by Water

Feature	Where to see a good example
Alluvial fan	See Alluvial fan, pp. 217–18
Arch, natural	See Natural arch and natural bridge
Bridge, natural	See Natural arch and natural bridge
Cave	Sierra Nevada district More than 100 caves; Moaning, Cave City, and Mercer's in Calaveras County are commercial caves open to the public; 11 caves are in Sequoia National Park (some are open to visitors), others in Packsaddle and Kern River Canyon Trinity–Shasta Lake district Shasta Caverns open to visitors Santa Cruz–Monterey district Caves badly vandalized Mojave Desert district Mitchell Caverns State Park open to visitors See also Lava tubes and Sea caves
Delta	Salton Sea Confluence of San Joaquin and Sacramento rivers near Suisun Bay
Marsh	Death Valley Suisun, Solano County Point Reyes National Seashore Humboldt Bay Various other locations along the California coast
Meander	Leavitt Meadow, Sierra Nevada

TABLE 8. Features Formed by Water (Contd.)

Feature	Where to see a good example
	Flat portions of major rivers throughout state
	Owens River near Bishop
Natural arch and natural bridge	Calaveras County
	Death Valley National Monument
	See also Sea arch, p. 177
River	See map 6
River capture (including water and wind gaps)	Alvord Mountains, Mojave Desert (imminent)
	Cajon Creek, Mojave Desert
	San Juan Capistrano
	Wheeler Ridge (wind gap)
	Pushawalla Canyon, Coachella Valley
	Furnace Creek at Zabriskie Point, Death Valley
Springs, hot or mineralized	Several in Warner Mountains and near Pit River, Modoc County
	North of Shasta, Siskiyou County
	Several in Lassen Volcanic National Park
	Near Amedee, Honey Lake
	Near Sierraville, Sierra County
	Grover's Hot Springs State Park, Alpine County
	Many in Clear Lake district, Lake County
	Casa Diablo Hot Springs, Mono County
	Near Bodie, Benton, and Bridgeport in Mono County
	In High Sierra near Ritter Range
	Hot Creek hot springs, Mono County

TABLE 8. Features Formed by Water (Contd.)

Feature	Where to see a good example
	Coso Hot Springs, Owens Valley
	Dirty Socks hot springs, Owens Valley
	Near Ballarat
	Death Valley National Monument
	Warm Springs, Alameda County
	Near Mt. Diablo, Alameda County
	In Santa Lucia Mountains, Monterey County
	Near Big Sur, Monterey County
	Near Paso Robles
	Along Kern River
	Ventura County
	Near Beaumont, Riverside County
	Near San Jacinto
	Murietta Hot Springs
	Near Salton Sea
	Zzyzx Hot Springs, San Bernardino County
Tafoni (honeycomb weathering, alveolar weathering, cavernous weathering)	Approximately 10 kilometers inland from coast near Pt. Conception
	Arroyo de los Frijoles, near Pigeon Point
	Mt. Diablo (Indian caves)
	Salt Point
	Death Valley National Monument
	Torrey Pines Beach, San Diego County
	San Mateo County (sandstone caves)
	Vasquez Rock, Agua Dulce Canyon, Los Angeles County
	10 kilometers west of Salsberry Pass, Highway 178
	Elephant Rocks, Marin County
	Castle Rock park

TABLE 8. Features Formed by Water (Contd.)

Feature	Where to see a good example
	Montara Mountain
	Death Valley
	Alabama Hills, Inyo County
Terrace, lake (former lake level)	Artists Drive, Death Valley National Monument
	Owens Lake
	Mono Lake
	Searles Lake
	Death Valley
	Ancient Lake Cahuilla, Salton Sea
	Ancient Lake Manix, near Afton, San Bernardino County (2 miles south of Interstate Highway 5)
	Panamint Range near Ballarat, Inyo County
	Near Hot Springs Mineral Spa, Imperial County
Terrace, river	Ventura River valley, Ventura
	Santa Ynez River near State Highway 154 at Armour Ranch road, Santa Barbara County
	Santa Clara River near ValVerde Park turnoff, State Highway 126, Santa Barbara County
	Santa Paula Canyon, near Santa Paula
	Interstate 5 north of Yreka
	Hungry Valley, near Gorman
	North of Bishop along U.S. Highway 395
	Silurian Hills, north of Baker

TABLE 8. Features Formed by Water (Contd.)

Feature	Where to see a good example
Tufa cones (tufa domes)	Mono Lake
	Honey Lake
	Searles Lake
Waterfall	
Beach	Pt. Reyes National Seashore
	Dana Point, Orange County
Dry	Little Lake, Inyo County
	Painted Canyon, Orocopia Mountains
Glacial hanging waterfall	Yosemite Valley, Yosemite National Park
	Hetch-Hetchy Valley, Yosemite National Park
Granite	Trails in Yosemite, Kings Canyon, and Sequoia National Parks
	Trinity Alps
Lava	Burney Falls
	Kings, Lassen Volcanic National Park
	Rainbow, Devils Postpile National Monument
	Horseshoe and Twin lakes near Mammoth Lakes

TABLE 9. Landslides

Feature	Where to see a good example
Landslide (including earthflow, mudslide, mudflow)	Puente Hills
	Gilroy slide, Gilroy
	Squaw Rock, Mendocino County
	Palos Verdes Hills, Los Angeles County
	Pacific Palisades
	American Canyon, Interstate Highway 80 near Vallejo
	Blackhawk slide, near Mt. San Gorgonio
	Death Valley Junction
	Palisades Cliff near Temescal Canyon
	Pacific Coast Highway near Topanga Canyon
	State Highway 33 near Adobe Creek, just south of Pine Mountain, Santa Barbara County
	Foster slide, State Highway 74 and San Juan Creek Road, Orange County
	Martinez Mountain slide west of Valerie Jean, Riverside County
	China slide
	Gorman slide
	Pt. Fermin, near San Pedro
	San Antonio Canyon, San Gabriel Mountains
	Orinda, Contra Costa County
	Mormon Temple district, Oakland
	Wrightwood, Los Angeles County
	Devils Slide, San Mateo County
	Pleasanton Ridge, Alameda County
	Marin County, along U.S. 101 and State Highway and throughout north Coast Ranges
	Death Valley National Monument

TABLE 9. Landslides (Contd.)

Feature	Where to see a good example
Rockfall	Yosemite National Park (fall of Eagle Rock)
	Tioga Pass
	Many other places in Sierra Nevada
Rockslide	Emerald Bay, Lake Tahoe
	Salton Sea
	Martinez, near Palm Desert
	Chaos Crags, Lassen Volcanic National Park
	Rose Valley, Coso Range, Inyo County
	Yosemite National Park
	San Jacinto Mountains

TABLE 10. Curious Features

Feature	Where to see a good example
Badlands (man-made)	Malakoff State Historic Park and many former hydraulic mines in Sierra Nevada and Klamath Mountains
	See also table 14
Concretion	Punkin Patch, near Tule Wash, Anza-Borrego Desert State Park
	San Felipe Hills
	Bolinas, at tide level
	Cannonball Wash, Colorado Desert
	Del Puerto Canyon, Stanislaus County
Dredge tailing (man-made)	South Fork, Scott River
	Brownsville
	Sacramento, near American River
	Snelling
	La Grange
	Atolia, San Bernardino County
Dragon's teeth	West shore of Searles Lake
Fantastic weathering	Pinnacles, U.S. Highway 395, 6 miles north of Alturas
	Hoodoos, Interstate Highway 80 near Truckee
	Death Valley National Monument
	Red Rock Canyon State Park
Petrified tree	Petrified Forest, Sonoma County
Tar seeps	Santa Barbara County near Carpinteria
	McKittrick
	Rancho La Brea, Los Angeles
Tombstone rock	Sierra Nevada gold country, especially along State Highways 4 and 49

6 • RIVERS OF ICE— RESHAPING THE LANDSCAPE

Glaciers are like the mills of the gods, in that they grind slowly, and can grind exceeding fine. It is not easy to see these great giants move, but they do move, and as they do, they make vast changes in the landscape. Indeed, much of the high mountain scenery of California has been so altered by glaciers of yesteryear that it is hard to imagine how it once was.

The Pleistocene Epoch, or "Great Ice Age," began in California about 3 million years ago. At its height, about 60,000 to 20,000 years ago, most of the Sierra Nevada was covered with ice, except for a few wind-swept peaks, too high to capture snow, projecting above the icy mantle. The Sierras were not the only mountains in California to be glaciated. The two major volcanoes of the Cascade Range in California, Mounts Shasta and Lassen, were glacier clad, as were the White Mountains and small portions of the San Bernardino Mountains, the Coast Ranges, and the Klamath Mountains. Today, glaciers still lurk in a few corners of the High Sierra and on Mount Shasta, but they have largely melted from other peaks. Ice Age climate was colder and wetter than today's; exactly why, no one knows, although more than 50 reasons have been suggested, including, among others, tilting of Earth's axis, sun spots, and volcanic eruptions.

If you would like to imagine the Sierra Nevada as it must have looked 60,000 years ago, go to Juneau, Alaska. There, in the southeastern Alaskan ice cap, you will see mountain glaciers spreading from the mountain crest, pushing long fingers down across the land. If you go, you will be following in the footsteps of John Muir, for he, too, went to Alaska to visualize California of yesteryear.

There is, however, a major difference. California's glaciers were confined to the mountains and did not reach the sea. Glaciers formed on Snow Mountain, Black Butte, and Anthony peak, all inland at the boundary between the northern Coast Ranges and the Klamath Mountains. In southern California, the nearest glaciers to the ocean were in the high parts of the San Bernardino Mountains. For that reason, there are no fjords in California; but the high glaciers left many other telltale marks on the mountain landscape.

The glaciers filled the mountain valleys (hence the name, "mountain," or "valley" glacier). They existed at the same time as, but were not connected to, the vast ice sheets that spread over Canada and much of eastern and northern United States. In the eastern part of California and other parts of the Southwest, lakes (called "pluvial" because they formed in those rainy days) filled much of the landscape. Large lakes occupied much of the Central Valley, and one lake, dammed by landslides or by movement on the San Andreas fault, stood east of Hollister.

Along the mountain crest, Mount Whitney and other high peaks projected as islands above a sea of ice—barren, wind-whipped deserts standing above the whited slopes. Below, ice filled the valleys, in places to the brim, and overflowed onto the slopes.

Despite the intense cold of those Ice Age winters of long ago, the mountain glaciers were temperate glaciers, similar to today's southeastern Alaskan ones, not nearly so cold as the polar giants of yesterday and today.

California's glaciers were, of course, much smaller than ice sheets of yesterday. Although the ice cap that covered the Sierra Nevada was more than 160 kilometers (100

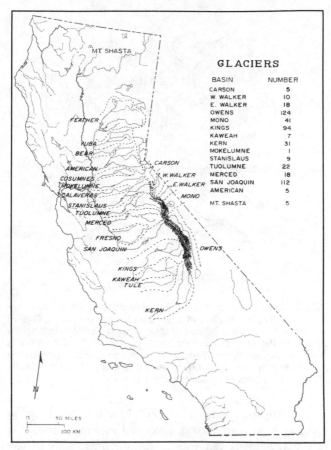

MAP 8. Living glaciers of California.

miles) long, 65 kilometers (40 miles) wide, and as much as 1,200 meters (4,000 feet) deep, this was a small patch compared to the giants that covered North America in those days, or to today's thirteen-million-square-kilometer (six-million-square-mile) Antarctic ice cover. California's longest glacier, the Tuolumne, was about 85 kilometers (60 miles) long.

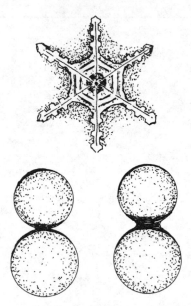

FIG. 28. How snowflakes, after becoming tiny spheres, metamorphose into glacial ice. Here at a constant temperature of −5°C (0°F) are two spheres of ice 0.5 mm (0.02 inches) in diameter. In the view on the left, the spheres have just made contact; on the right, 5 hours and 20 minutes later, a discernible neck has formed between the two. Eventually, if cold continues, they and some of their neighbors will become a single crystal of glacial ice.

The progress from gentle snowflake of intricate design to grinding blue glacial ice is a story of metamorphism. It is rare that humans can witness the metamorphosis of one rock into another, as the progress often takes place deep within the heated bowels of the earth at a pace dictated by geologic time—far longer than our brief days. With ice, however, we can actually watch the change.

First, the loosely packed snowflakes change into very small spheres. Where the spheres touch, little necks form, connecting them. Water vapor from the tiny balls migrates to the necks to form larger and longer spheres as the entire mass welds together. All of this can take place below the freezing point, so water is not necessarily part of the change. If the temperature rises above freezing so as to melt some of the snow, the process speeds.

As the ice mass melds together, it increases in weight and strength. Fresh snowflakes weigh about a quarter of a gram per centimeter—about a tenth the weight of water. In a week or two, they have become powder snow, twice as heavy and twice as strong as snowflakes. In months, they become "old snow," doubling again in weight and strength. In years, they become "firn," the material of which glaciers are made. In hundreds of years, they have consolidated into glacial ice, a form of rock nine-tenths as heavy as water. During this metamorphic process, they have increased in weight 9 times and in strength 500 times, and have altered their appearance from dainty filagrees to dense crystals of glacial ice as long as 25 centimeters (10 inches).

As a stream of water is usually deepest in the center and shallowest at the edges, so glaciers (streams of ice) are thickest in the center; but the surface of a stream of ice, unlike a stream of water, is usually bowed upward in the center. For this reason, water melting on the surface of a glacier will run toward the sides as well as down slope, causing glacial streams to converge toward the sides of the glacier rather than in the center, as liquid streams do.

Most mountain glaciers have a large crack or "berg-schrund" a few feet from their upper (headwall) end, and parallel to it. Sierran glaciers of today—Dana, Conness, Lyell, Palisade glaciers—each have one.

As a glacier moves downhill, other cracks, called "crevasses" form in its body. Crevasses may be less than a centimeter or more than 15 meters (50 feet) wide; they may be a few meters or hundreds of meters long. They form where the glacier is under stress: where it turns a corner, where it plunges over an ice fall, where there are projections in the valley floor, or in steep stretches where the central ice is moving faster than the ice along the edge. Each of these cracks, including the bergschrund, is a place where rocks and dust can lodge.

The environment of ice in a mountain glacier is somewhat similar to that of a river, yet different in significant ways. There are streams within the ice, some along the edge, some within the glacier body. They behave like streams everywhere in that they flow generally downhill and carry objects, largely sand and gravel, in their beds. Unlike water on land, however, these streams, which are confined to tubes of their own making in the ice body, can occasionally flow uphill. So, also, can glacier ice. If it encounters an obstacle in its generally downward path, it can ride up and over it, often shaving, changing, planing, remaking the object in the process.

Glaciers, like rivers, follow the easiest course; for that reason, California's glaciers chose paths already made for them by previously flowing streams. Unlike rivers, glaciers do not start from fairly undefined beginnings, and flow downhill in a dispersed manner to merge themselves into the sea. Mountain glaciers are more like lakes, in that they have a definite beginning (the head, or cirque, end) and an abrupt end (the snout). Even glaciers that feed into the sea

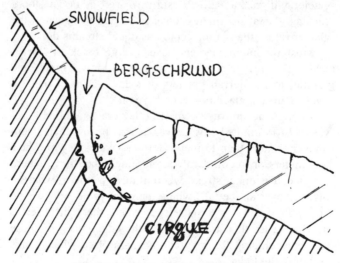

FIG. 29. Diagram of the head of a glacier.

FIG. 30. Glacially striated rock.

have a marked ice cliff at the water's edge, although they may extend under water.

Both ice and water move downhill in response to gravity. Water moves quickly; ice slowly. Glaciers do occasionally surge, or "gallop"; one was clocked recently at 100 meters (300 feet) per day.

Some of the glacier movement is due to melting and refreezing. When the glacier reaches about 30 to 50 meters (100 to 150 feet) in thickness, its weight can cause it to move molecule by molecule within itself.

Because a glacier does move, it changes the landscape in unusual ways. By freezing to the enclosing rock, it can "pluck" or "quarry" huge blocks of rock from the enclosing sides and bed. These and many smaller blocks, as well as gravel, sand, and dust, fall on the glacier and are swept into its crevasses or are carried along its bottom. There, mixed with water, they form an immense file that scours and reshapes the glacial valley. The file, too, is worn down as the glacier progresses, grinding rock and sand into glacial dust, or "flour," and polishing the blocks it is using.

When a glacier has melted from its valley home, the alterations it has made are exposed for us to see. At its head, high in the mountains, a glacier carves an amphitheater-shaped depression called a cirque. After the ice has gone, the cirque is often occupied by a clear, cold mountain lake called a tarn. From the cirque, the glacier may have moved in a series of steps down the mountain; when the ice has gone, each of these steps may be occupied by a glacial lake, connected to the one below it by a glistening mountain stream. From a distance, these lakes reminded early mountaineers of beads on a silver thread; from this, they take one of their names, "paternoster" ("our Father") lakes, because they resemble beads on a rosary. They are also called glacial step lakes, cyclopean stairs, giant's staircase, and glacial stairways.

Ice may fill an entire valley and be joined by other tributary ice streams, which may meet the mainstream of ice at the same level, or drop into it as ice falls. When the ice has melted, it may be apparent that the bottoms of the main valley and the tributary valleys are not at the same level, although their icy tops may have joined. When this is true, the tributary valley is left as a high, hanging valley that may, in wet years, contain a stream that plunges from the tributary valley into the main stream over a breathtaking waterfall. Yosemite Valley has several such waterfalls: Vernal, Bridalveil, Yosemite, and Nevada Falls, among others.

The valleys themselves are changed by the glacier. Where once each valley may have had a narrow V-shape cross section, passage of a thick glacial tongue has worn it to a U. This has been accomplished by scraping and gouging with the rocky file; by freezing and thawing of the ice in which the rock is caught. In the process of remodeling the valley, the glacial file may leave telltale marks on the rock of the valley walls: gouges, grooves, scratches, glacial polish, chatter marks. If there had been rocky projections in the glacier's path, the glacier may have worn them to a strung-out, lumpy shape called "crag-and-tail" or "roches moutonneés."

This latter name, French for "rock sheep," comes from the resemblance of the rock to fancy wigs, worn by eighteenth-century gentlemen, which were dressed with

A miniature California landform: this 2-foot-high salt pillar, located on a Death Valley playa, has been worn by rain and trimmed by wind.

Shapes of volcanic landforms: photos (above) show the linear structure of the volcanic piles at Devils Postpile National Monument, a dark-colored lava flow (top surface is on the bottom right of photo); an eroded cinder cone (left) built of sand-sized volcanic particles; rounded dome (below) of one of the Mono Craters, formed by cooling of sticky, light-colored lava.

Until the eruption of Mount St. Helens in 1980, Mount Lassen (right) was the only volcano in the contiguous 48 states to have erupted in the twentieth century. On May 22, 1915, the day of Lassen's most vigorous and spectacular eruption, the mountain threw a 30,000-foot-high mushroom cloud into the air, as well as a hot blast laterally down the mountain that destroyed trees as far as three miles away, resulting in a "devastated area" of more than a square mile where no trees remained standing. Mount Lassen's eruptions were much milder than Mount St. Helens in 1980, however. Mount St. Helens created a devastated area nearly 16 miles long that covered 150 square miles. But what is devastation today may be beauty tomorrow. The photograph (bottom) taken at Lost Lake, Alpine County, shows volcanic country many thousands of years after eruptions have ceased.

Sheeting in granite, Yosemite National Park (left). Although some molten rock is poured or blown out over the surface of the Earth through volcanoes, a great deal of once-molten rock forms the cores of such vast mountain ranges as the Sierra Nevada. Here, the molten rock has solidified more slowly deep underground, forming such familiar rocks as those of the granite family (left and bottom). The processes of erosion have since unroofed the now solidified mass. Granitic rocks can be seen throughout the Sierra Nevada and also in northern California in the geologically related Klamath Mountains (bottom right). The southern California ranges are also composed of a related granitic group that extends into Mexico. Some of California's mountains have been lifted by fault action, such as the Sierra Nevada, or have been moved many miles laterally along faults. In the upper view on this page, the infamous San Andreas fault cuts through the Coast Ranges, creating a valley now filled with sag ponds. Shown here are Crystal Springs Reservoir and San Andreas Lake, from which the fault takes its name. A chamois prepares to leap from a sandstone crag in San Diego's Wild Animal Park (below).

Ice age relics in today's California desert. These are scenes of Mono Lake, a relict lake (top right and bottom). During those far-off days, ancestral Lake Mono was much deeper and larger than it is today. It was part of a system of large lakes in eastern California which had its final termination in Death Valley. Today, ancient strand lines surround the lake like rings on a giant bathtub. The level of today's Mono Lake, the only natural lake left in this chain, has gradually fallen as the climate has dried, revealing tufa domes (top right), limestone structures that mark hot springs or colonies of algae in the lake. Today, the domes are a fairyland of white towers beside the blue waters of the lake. Mono Lake is dwindling very fast as the city of Los Angeles diverts most of the water that would normally feed it. Already thousands of birds that commonly nested on the two islands in the lake have become prey to marauders who have access to the islands as the waters recede, leaving mudflats as bridges connecting mainland and islands. If Los

Angeles continues to divert the water, Mono Lake, called "the Dead Sea of California," will soon be as dry as Owens Lake, farther south, which was dried up in the early decades of this century, its drying process speeded by water tapped for Los Angeles as well. Perhaps all that will be left someday of Mono Lake will be an alkaline pond similar to the salt pond in Death Valley (left). The lake floor will then provide ample fine particles to feed potentially dangerous dust storms.

The ferocity of dust storms is lessened in parts of the desert by a protective armor of stones called "desert pavement" (right), which prevents the fine clay particles beneath the stones from being lifted by the wind. When the covering is damaged, as it is by off-road vehicles, the underlying fine particles are exposed, and can become a health hazard to neighboring communities when they are blown by the wind.

When there is no cover of stones, desert clay cracks into a huge mosaic as the sun dries mud after brief storms (left).

Wind can work forcefully in the desert where little moisture and few plants hold down the soil. Although one thinks of deserts as being sweeping vistas of sand as in the part of Death Valley National Monument shown above, a relatively small portion of the world's deserts is covered by dunes. As the wind piles and moves sand, it makes patterns of many sizes and shapes, ranging from small ripples on the sides of single dunes (left) that shift with each passing gust, to huge sand seas whose pattern can be discerned by satellites high above Earth. Most of the particles moved by the wind have been dislodged from their source rock by water or ice and are merely being moved by the wind. Rarely can the wind erode rock itself; when it does, the worn portions are seldom higher than a few feet above the ground.

This photograph, taken in Joshua Tree National Monument, shows a group of rocks with "waistlines" cut by the wind.

Although a host of tiny glaciers still hangs in the Sierra Nevada, none is larger than a square mile in area. Even the largest is minute beside the giants of the Great Ice Age, melted these thousands of years, and is dwarfed by even the tracks of those ancient glaciers. In this photograph (top), moraines of rocky rubble at Laurel Canyon (on the left in the photo) and Sherwin Canyon (on the right in the photo) form hills in the foreground. The grayer moraine is made largely of metamorphic rock and supports only

sage; the higher, greener moraine is composed largely of granitic rock and supports evergreen trees. The present course of Laurel Creek is a sinuous yellow line in the photograph, made by the fall color of aspen and willow trees growing on the banks of the creek.

Besides leaving heaps of rubble, the old glaciers polished the rock over which they rode (right, above, taken at Devils Postpile National Monument), and cut basins for high mountain tarns, as at Canyon Creek in the Trinity Alps (lower left).

The highest peak in the contiguous 48 states is Mount Whitney, here shown towering above the Alabama Hills near Lone Pine. The sharp, steep eastern face of the Sierra Nevada has been lifted by faulting and sharpened by glaciers. Although the peaks are made of the same kind and same age of rock as the granitic rock of the Alabama Hills (foreground), the difference in aspect is striking. While the high mountains were being glaciated, the rock of the Alabama Hills was buried. The wetter days of the Great Ice Age provided more snow to the mountains, and much more water to the lower regions, so that the rocks of the Alabama Hills began to rot underground. Erosion has since stripped them and exposed them to the desert sun and wind (see also cover photo).

The battle between sea and land is constant, but all else is constantly changing. Erosion works so rapidly at the sea shore that it transforms the coast before our very eyes. Here are two pairs of photographs showing changes that have taken place within a score of years. One of the sea arches (top) at Natural Bridges State Park, Santa Cruz, taken in 1978; the same rock in 1981 (bottom left), but the arch has fallen.

On the page to the right is a sea stack as it appeared in 1978; on the left is the same sea stack as it had appeared twenty years earlier when the protecting crown was much larger. If one multiplies this amount of change by the eons of time that have passed to make California's landscape what it is today, one can easily see how vast alterations are possible.

A sea stack at Santa Cruz.

The coast changes even as we watch. The pounding surf literally explodes against the cliffs, grinding them to pebbles, and the pebbles to sand. Winter seas are far stronger than summer seas, as they bring heavy storms to beat upon the shore. During such storms, the waves can transport huge boulders. The winter beach is often littered with cobbles (bottom right). In contrast, during the summer, the sea can handle only smaller particles of comfortable sand on which to bask in the warm sun, as is this elephant seal at Año Nuevo (top).

There are few places, if any, left in California "where the hand of man has never set foot." One of the first changes humans made in the landscape was to fill in the edges of bays with debris to make more dry land. In this photo, the marshes have been diked in order to harvest salt. Each pond is a different hue because each is a different salinity in its progression from seawater to solid salt, and can thus support a different species of alga, which tints the salty water.

FIG. 31. How a glacier can change a stream valley. *Top*, The land before glaciation has smooth, rounded hills with streams wandering through. *Center*, Glaciers cover the landscape, steepening the ridges and peaks of the high country. *Bottom*, The glacier has melted, revealing the new landscape that has been carved by the ice.

sheep tallow. Since it is also true that from a distance a group of such glacier-worn rocks resembles a herd of sheep, the geologist who gave them this name probably intended a double pun. The tail, or strung-out portion, betrays the direction from which the glacier came, smoothing the rock as it did so. The "crag," or downstream portion, is rough and hackly, as that part was eroded by the freeze–thaw of glacial plucking.

A geologist who studied mountain glaciers has compared the topography left after melting of the ice to the dough left on a biscuit board after the biscuits have been cut and removed. One can see two adjacent cirques, where once the heads of two glaciers nearly met, that have now become empty sockets with a knife-edged ridge (arête) between; where the heads of three glaciers nearly met, a sharp peak, shaped like a horn, is now left. The name "horn" or "matterhorn" is given to this alpine feature, after the Matterhorn, in the Swiss Alps, which geologists think was carved by glacial ice in this way.

Glaciers build landscapes as well as erode them. Along its sides, where streams ran, and where debris has fallen from the surrounding cliffs, the melted glacier leaves a rounded ridge of miscellaneous rock of all sizes and shapes. Other ridges of the same heterogeneous material are formed at the end, where the glacier has dumped them like a conveyor belt. If the glacier advances, it pushes up the ridges, much as a snowplow pushes snow. In some places several successive ridges can be seen at the end of a glacial valley, marking different halting places of the glacier. Because ridges marking a glacial retreat can be erased by a succeeding advance, it may be difficult to decipher the entire history of a glacier. In the final retreat, however, the debris farthest downhill is the oldest.

These lateral and terminal piles of debris are called by the French word, "moraine," meaning hill. After a glacier melts, the moraines are left behind to delineate edges and ends; where two glaciers join, a moraine may form between them. If one can determine the age of each moraine, it is

FIG. 32. How a glacier leaves heaps of rock debris as it melts back. Along the sides are sharp-crested ridges of rock (lateral moraines) curving around the end to form a series of terminal moraines. Each inner loop is younger than the outer one, as it marks the position of a retreating glacier. In the right foreground, a glacial meltwater stream has cut through the terminal moraines, carrying away some of the the morainal material.

possible to trace the history of the glacier. Some moraines in California contain fragments of wood whose age has been determined; others have been covered by lava flows whose age in years is known. Using these methods, geologists have worked out the chronology of some of California's glaciers in detail.

Rock and sand that fell on top of glacial ice and were still riding there, together with material carried by the streams within the ice, drop to the ground as the glacier melts. In this way glaciers carry boulders many miles from their source. These boulders, which are strangers in the landscape where they are left, are called "erratics"—perhaps they are chunks of granite in limestone country or perched incongruously on sandstone bedrock.

Here and there, one can find a boulder in glaciated country that has one or more flat faces, similar to wind-worn

FIG. 33. Erratic, a rock transported by a glacier that is now out of place in the landscape. Many erratics are of a different rock type than the rock they are resting on.

pebbles of the desert. These have been tools of the glacier, used for scraping against the walls and bottom of its bed. With such tools as this, the glacier has sculpted the landscape into the mountain grandeur of today.

Yosemite National Park is probably the best place in the world to see what glaciers can do to a mountain landscape. If you stand at Glacier Point and look westward, you can see how the valley has been changed. Here, you can tell where two ice streams have joined (just below Glacier Point) to merge as one ice stream to carve Yosemite Valley. As the ice moved past Half Dome, it cleaved off the dome face along the joints, carrying away the rubble to leave a clean, steep precipice. At the point where you are standing the glacier completely filled the valley with ice. The side valleys feeding into it were full, as well. In Ice Age days, they joined into one thick stream. Now that the glaciers have melted, you can see the disparity between the levels of the bottoms of the canyons of the main and tributary streams, for they are marked by waterfalls plunging hundreds of feet into the valley.

From this vantage point, too, you can see large and small roches moutonneés. Liberty Cap and Mount Broderick are two surviving roches moutonneés; they were originally rock outcrops that stood directly in the path of the glacier, and were reshaped, but not destroyed, by it. Most other domes in the park are not actual roches moutonneés, as they have not been overridden by ice, but have been altered or polished by its passage.

Many moraines dot Yosemite Park, but it is easier to see those on the eastern side of the Sierra Nevada because fewer tall trees mask them there. Long, curved morainal ridges surround Convict Lake, Walker Lake, Twin Lakes, and many of the smaller lakes that cluster along the eastern foothills. Some of these moraines are quite high; those, for example, near Mono Lake are about 250 meters (800 feet) high.

The many small glaciers that lie high in California's mountains today are more curiosities than sculptors of the landscape. All of them are making some slight impression, but with the exception of the glacier on Mount Shasta that caused mudflows that coursed down the mountain in the 1920s, none are producing changes that compare with the changes made by the giants of yesteryear.

These tiny glaciers are not remnants of the mighty giants of the Great Ice Age. The last of the great glaciers melted completely about 10,000 years ago. These new glacierets have formed just lately—geologically speaking. They take their origin from the cold days of the Little Ice Age (called the Matthes glaciation after François Matthes, a California geologist who discovered it). The Little Ice Age glaciers were at their height from about A.D. 1700 to 1750, during a time when the climate in North America was much colder.

In an aerial survey of Sierra Nevada glaciers in 1972, the U.S. Geological Survey counted 497 glaciers and 847 ice patches. All are small, the largest being Palisade Glacier, near Mount Whitney, which has a total area of 1.6 square kilometers (0.6 mile), and a length of 1.9 kilometers (1.1 miles). Compare this with Alaska's Hubbard Glacier, which is 75 miles long and several miles wide, and Seward Glacier, which is 40 miles long. All of California's 497 glaciers together have a total area of 51 square kilometers

FIG. 34. Konwokiton Glacier, Mount Shasta, as it appeared in 1896. Konwokiton is one of five glaciers presently mantling Mount Shasta.

(19.7 square miles). Add to this the area of the ice patches —14 square kilometers (5.4 square miles)—and the total Sierran ice adds up to only about 65 square kilometers (25 square miles). Although this is insignificant as glaciers go, it is an important contribution to our water supply.

Once in a while, sudden melting of ice in a glacier may cause a jökulhlaup, an Icelandic word for "glacial outburst flood." Such floods are sometimes caused by fresh, hot lava from volcanoes.

In California, Mount Shasta's Konwakiton Glacier triggered jökulhlaups in 1881, 1920, 1924, 1926, and 1931, as well as many prehistoric ones. The 1924 outburst, which was witnessed by campers and workers, blocked roads and destroyed the water supply for the town of McCloud, and sent yellow volcanic mud out into the McCloud and Sacramento rivers. Unlike Icelandic outbursts, Shasta's jökulhlaups have been blamed on the heat of the Sun, rather than volcanic eruptions, although it is always possible that the old cone may someday cause a flood by its own internal heating plant.

What are the prospects for another ice age? Most scientists agree that we are probably in a stage between worldwide glaciations. Some think that we are already on the coldward path, and that we may reach another glacial stage relatively quickly, geologically speaking. One grave danger (unless we desire another glacial age) is that pollution of the air from various causes, including volcanic eruptions, supersonic planes, aerosols, smoke, and dust, among others, may trigger a series of reactions that could lower the mean temperature of Earth by a few degrees. A drop of only 2°C could trigger a new ice age. Or, conversely, if the temperature were to rise for any reason, the polar ice caps could melt, thus raising the sea level and flooding seacoast cities all over the world.

If a new ice age or ice-caused flood does come, it will probably happen slowly enough that none of us alive today will see the landscape altered by glaciers or glacial floods of tomorrow.

TABLE 11. Glacial landforms of California

Feature	How to recognize it	Probable origin	Where to see a good example
Arête	Steep, sharp rock ridge between adjacent glaciers	Quarrying by glacier in cirque	North Palisade
			Mt. Humphreys
			Kaweah Crest
			Ritter Range
			Mt. McClure
			Mt. Lyell
			Septum between Mt. Whitney and Mt. Russell
			South of Mt. LeConte
			(All in Sierra Nevada)
Avalanche chute	Slick, steep, U-shaped groove barren of vegetation	Snow avalanche, following a particular path	Bearpaw Meadows, High Sierra Trail
			Sequoia National Park
			Mt. Whitney
			Mt. Hitchcock
			Hamilton Lakes
			(All in Sierra Nevada)
Chain lakes	See table 7		
Chatter mark	Crescentic gouge or fracture, deeper on downstream (down-ice) end	Impact pressure of ice	Mt. Huxley
			Evolution Basin, near Sapphire Lake
			Sierra Nevada

TABLE 11. Glacial landforms of California (Contd.)

Feature	How to recognize it	Probable origin	Where to see a good example
Cirque	Bowl-shaped depression in mountain-side, gener-ally backed by steep cliff	Scouring of glacial ice at head of glacier; plucking of headwall	High country of Sierra Nevada Trinity Alps Thompson Peak Snow Peak near Clear Lake
Cirque lake	See table 7		
Col (pass)	Low saddle in glacial ridge opposite two cirques	Erosion by the heads of two glaciers, co-alescing to destroy part of arête	Mono Pass, above Bloody Can-yon, Sierra Nevada
Erratic	Markedly dif-ferent type of rock lying in terrane that is not its source	Boulder car-ried on or in ice, left when ice melts	Chiquito Creek Faith Valley June Lake Rock Creek Balloon Dome Sentinel Dome Starr King Meadows Moraine Dome Cathedral Rocks Twin Lakes near Kaiser Peak Bighorn Plateau

TABLE 11. Glacial landforms of California (Contd.)

Feature	How to recognize it	Probable origin	Where to see a good example
			Cisco Butte, 2.5 kilometers (1.5 miles) west of Cisco Grove
			Loon Lake
			Yosemite National Park
			(All in Sierra Nevada)
Glacial outburst flood deposits	See Jökulhlaup		
Glacial polish	Shiny surface on rock, but shine may be worn off in spots, if weathered	Polishing by finely ground rock in glacier ice	Domes in Yosemite Park and throughout Sierran high country
Glacial stairway (glacial staircase, cyclopean stairs, giant's staircase, glacial steps)	Series of flattish valley areas, commonly with lakes connected to one another by steep areas often having water falls (See also table 7)	Scouring of glacier bed (at bottom of glacier)	Yosemite Valley
			Sixty Lake Basin
			Faith, Hope, and Charity valleys
			Big Pine Creek
			Pine Creek
			Black Rock Pass
			Piute Pass, near Bishop
			(All in Sierra Nevada)

TABLE 11. Glacial landforms of California (Contd.)

Feature	How to recognize it	Probable origin	Where to see a good example
Hanging valley	Tributary valley much higher than main valley, usually marked by steep cliff or perhaps a waterfall	Thinner tributary glacier met main glacier at level where tops were even, but bottoms were not; when ice melted, channel of tributary was left much higher	Yosemite Valley Hetch-Hetchy Valley Evolution Basin
Jökulhlaup (glacial outburst flood) deposit	A layer of sediment containing large and small rocks in a mud matrix	Sudden melting of glacier	Mud Creek, Mt. Shasta Lamarck Glacier, Mt. Goddard quadrangle
Kame terrace	Mound of poorly sorted sand and gravel forming ridge along glacier edge	Deposited in channels of streams at edge of ice	Junction of Sonora Pass Highway (State Highway 120) and U.S. Highway 395
Knife-edged ridge	See arête		
Matterhorn (horn)	Steep sharp rock or peak	Quarrying in cirques of three adjacent glaciers	Matterhorn Peak near Bridgeport Mt. Emerson near Bishop Crest of High Sierra

TABLE 11. Glacial Landforms of California (Contd.)

Feature	How to recognize it	Probable origin	Where to see a good example
Moulin (pot-hole, giant kettle)	Cylindrical holes in rock of glacier floor	Grinding of rock by boulder or pebble in eddies and vortices of water within ice	End of Tuolumne Meadows, Yosemite National Park
Moraine: terminal, lateral, and medial	Long distinctive ridges of unsorted sand, gravel, clay, and boulders; some boulders may be faceted	Glacial sediment deposited at sides (lateral), in center (medial), or at end (terminal) of glacier; medial most commonly forms where two glaciers join	Moraines enclose Walker Lake, at foot of Bloody Canyon McGee Canyon (Sierra Nevada) Sawmill Creek (Sierra Nevada) Lee Vining Canyon Twin Lakes, Matterhorn Peak quadrangle Lateral moraine near State Highway 168, Huntington Lake quadrangle Convict Lake, near Mammoth Lakes Lundy Canyon, near Mono Lake

TABLE 11. Glacial landforms of California (Contd.)

Feature	How to recognize it	Probable origin	Where to see a good example
			(All in Sierra Nevada)
			Kuna Glacier, Dana Glacier, Palisade Glacier, Lyell Glacier, Mt. Shasta Glaciers (all modern glaciers having small terminal moraines)
Moraine-dammed lake	See table 7		
Outwash plain	Stratified deposit in a valley beyond moraine	Deposited by glacial meltwater carrying fragments of eroded moraine	Sand Meadows (Sierra Nevada)
Paternoster lake	See table 7		
Perched boulder (glacial table) (See also glacial erratic)	Pedestal of local rock capped by erratic; may have a mushroom shape	Local rock protected from weathering by more durable erratic	Starr King Meadows Upper Yosemite Falls Moraine Dome Chiquito Creek Parker Creek (All in Sierra Nevada)

TABLE 11. Glacial landforms of California (Contd.)

Feature	How to recognize it	Probable origin	Where to see a good example
Roche moutonnée	Rocky outcrop in glacial landscape, rounded on one side, irregular on the other	Erosion by overriding ice, smoothing upstream side by abrasion, plucking downstream side	Tuolumne Meadows Liberty Cap Mt. Broderick Desolation Valley Yosemite Valley
Rock glacier	Corrugated mass of angular rock near cirque, shaped like glacier	Mixture of ice and rock, moving like glacier, in areas not now cold or wet enough for glacier to form	Eastern side of Sierra Nevada Sherwin Canyon Mt. Tom Cirque between Two Teats and San Joaquin Mountain Kaweah Basin
Striations, grooves	Long scratches or indentations in rocks	Abrasion of glacial walls and floor by rocks embedded in ice	Evolution Basin Yosemite Valley Bloody Canyon Hamilton Lake Kern Canyon Pine Creek Pass near Bishop

TABLE 11. Glacial landforms of California (Contd.)

Feature	How to recognize it	Probable origin	Where to see a good example
			Agnew Meadow, Devils Postpile National Monument
			Saddlebag Lake, Yosemite National Park
			(All in Sierra Nevada)
			Lassen Volcanic National Park
Tarn	See table 7		
Till	Jumbled mass of clay, sand, and boulders; some boulders may be as large as 25 feet in diameter and some faceted; distinguished from volcanic mud flow by presence of large numbers of nonvolcanic boulders	Deposited on sides or below glacial ice, or in moraines	Near Bridgeport
			Reversed Creek
			Convict Lake
			McGee Creek
			Rock Creek
			Pine Creek
			In Bloody Canyon
			In Sawmill Canyon
			Around Lake Tahoe
			Around Lake Mary

TABLE 11. Glacial landforms of California (Contd.)

Feature	How to recognize it	Probable origin	Where to see a good example
			(All on east side of Sierra Nevada)
			Yosemite National Park
			Most Sierran passes
			Mt. Shasta
U-shaped valley	Valley having horseshoe shape in cross section	Eroded by glacier, usually modifying stream valley	Yosemite Valley
			Hetch-Hetchy Valley
			Kern River Canyon
			Evolution Basin
			Pine Creek Canyon
			Lake Mildred
			Shepherd's Crest
			Up and down the High Sierra

7 • CHANGE THROUGH EARTH MOVEMENT

"California, with all thy faults, we love thee still" is an amusing play on words which I once misquoted as "California, *for* all thy faults, we love thee still." It was a Freudian slip. The truth is that, despite the danger of earthquakes, we should, indeed, love California *for* her faults; for her faults, and other related earth movements, have made her what she is today.

The word "tectonic" (from the Greek word for builder) is used to describe the processes of earth movement that have created mountains and valleys, seas and shores. Tectonic movements have formed the major elements of the landscape, and have guided the development of Earth's facial features.

Since 1965, the earth sciences have been in intellectual revolution. Our view of Earth's long history has changed radically. Although we knew the stories of many of its parts—the story, for example, of the Devils Postpile in the Sierra Nevada, of Ben Lomond in Scotland, or of the Dry Valleys of Antarctica—we had very little idea as to how these parts were related to one another. It was as if we knew characters and scenes from a play, but had only a vague idea of how the plot went.

Now, however, earth scientists have a unifying theory called "global tectonics" that reconstructs the "plot." This theory views the Earth as being composed of several layers.

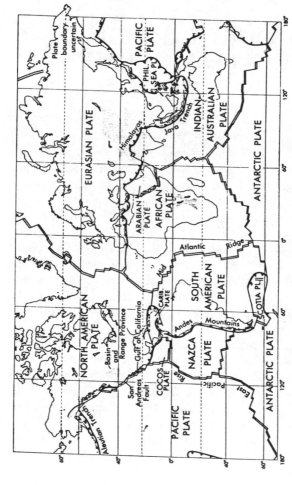

FIG. 35 The Earth's tectonic plates, showing active boundaries. Double lines show areas along which plates are moving apart; lines with barbs show zones where one plate is thrusting under another; straight lines are fault lines.

The outermost layer, called the crust, is from 5 to 55 kilometers (3 to 35 miles) thick, and consists of several large plates that carry the continents and the seas. These plates move about, sliding, more or less, on Earth's next underlying layer, the 3,000-kilometer (1800-mile) thick mantle. As the plates move, they may collide, forming mountains, slide by one another, or rip apart much as cloth is torn.

About 200 million years ago, it is postulated, the Earth's land was one large continent which gradually broke apart—"drifted"—into separate pieces. The joins of the last breakage are the edges of today's continents; looking at the neat fit, one can see where South America has ripped from Africa, and North America from Europe.

Proof that this has actually taken place has come from a number of sources. For one thing, the continental pieces do fit together remarkably well, if one imagines them joined at the lip of the continental shelf. In addition, animals, plants, fossils, and rocks in continents now separated by ocean are remarkably similar; and there is evidence that climates have changed drastically on some continents (as shown by tropical fossils found in rocks now near today's poles).

This and other evidence indicate that the continents have pulled apart. Indeed, the sea floor itself appears to be "spreading." The past few decades have allowed us for the first time to glimpse what is in the ocean depths. What we have seen by instrument, and, in some cases, by actual visits to the bottom of the sea, is that the sea floor is made up of a series of basalt (lava) stripes, arranged in a geometric pattern. By measuring the ages of the stripes using radiometric methods, and by determining their polarity (that is, by measuring the orientation of tiny magnetic minerals within them), scientists have discovered that the sea floor is growing from the center by the eruption of new lava, while the sea is spreading toward the edges. Since Earth is finite, seas cannot spread forever; to balance this spreading, the plate tectonic concept (or "model") suggests that this spreading is being balanced when the edges of the plates are consumed in a "subduction" zone, where the old edges are pulled back under the Earth's crust into the molten mantle.

Earth movements do not always break the Earth. At times, rocks will bend without breaking, forming great bends or "folds." (Folds that are bent upward like a croquet wicket are called "anticlines"; those that bend downward into a "U" are called "synclines.") Earth tectonic movements have pressed some mountain ranges, such as the Appalachians, into an intricate series of crests and dips (anticlines and synclines) which were later eroded to form mountains and valleys. Although most California mountain ranges have folds somewhere in their architecture, none of them is a wholly folded range. California is so intricately laced with faults that an unbroken fold is a rarity.

This global tectonic view of Earth is a very elegant one in a mathematical sense; that is, it fits the pieces of evidence

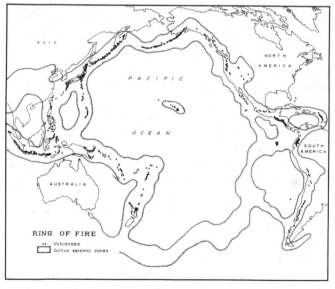

MAP 9. The Pacific "Ring of Fire" showing active volcanoes and active seismic zones.

MAP 10. Major faults of California.

together with few loose ends. But loose ends remain. A wise professor of mine once said, "What makes it happen? What force is strong enough to move continents?" We do not yet know the answer.

The San Andreas fault, the most famous fault in the world, separates two major Earth plates, the Pacific and the North American. Because the fault is a major plate boundary, much tectonic activity takes place along and near it. Earthquakes are frequent here; most of them are small, but now and then a major one occurs, and, infrequently, a great earthquake, capable of dealing death and destruction, changes the landscape.

Scientists are not entirely in agreement as to how far the San Andreas fault extends, but the most recently published fault map of the state shows it to run from the Salton Sea to Cape Mendocino. Perhaps it extends farther south into Mexico and farther north beneath the sea. I say "it," but "it" is not apt, because the San Andreas fault is not a single break, but rather a zone with breaks grouped together like an unkempt skein of yarn. In any earthquake, one or several of the fault strands may be active; in the next earthquake in the zone, a different strand may be the culprit.

Permanent change of the Earth along these fault strands is not vast in any one earthquake. Judging by landscape evidence, most of the San Andreas movement has been horizontal; the greatest in historic time was 21 feet of horizontal movement in the great San Francisco earthquake of 1906, measured at Olema. This earthquake was one of the world's greatest; compare the 21-foot horizontal change in

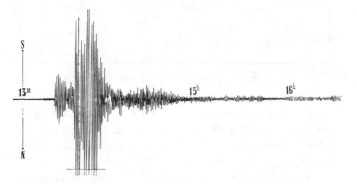

FIG. 36. Seismogram of the April 18, 1906, San Francisco earthquake as recorded in Siberia.

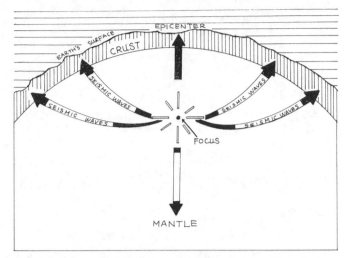

FIG. 37. Seismic waves start at a point called the "focus," deep within the Earth. They proceed to the Earth's surface where they can be detected by seismographs. The point on the surface directly above the focus is called the "epicenter." California earthquakes have shallow foci, generally less than 15 kilometers (10 miles) below the epicenters.

this 'quake with an 8-foot vertical movement in the much smaller "moderate" San Fernando earthquake of 1971, having a magnitude of 6.5. Although the movement in one was horizontal, and in the other vertical, both were Earth movements.

The magnitude of an earthquake is a mathematical measure of its power, while "intensity" measures the possibility and extent of damage to human structures. Both measure a single event; neither considers the geological effects, which, to be appreciated, must be viewed from the perspective of time.

Through time, the San Andreas, which is presently "creeping" (that is, moving by slow, nearly imperceptible spurts), has translated parts of southern California northward. For example, land now at Point Reyes was torn from

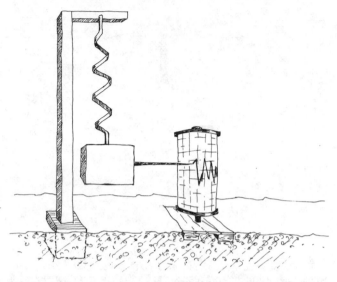

FIG. 38. How a simple seismograph works. A weight, suspended on a spring, moves with the earthquake to make a tracing on a paper. This seismograph is oriented to respond only to shaking in a vertical direction; other instruments must be used to record earthquake movement in the two horizontal directions.

the Santa Lucia mountain block and pushed along the plate boundaries to its present position. In southern California, parts of the Orocopia Mountains seem to belong to the Transverse Ranges, indicating that the southland, too, is shifting. It has been calculated that, in a few million years, Los Angeles and San Francisco will be side by side, if they still exist.

In some places, "creep" is shifting the blocks at a rate of about 2 centimeters per year; in other places, the fault appears to be locked. It is by no means certain that large earthquakes will not occur where there is creep; although creep is surely relieving some of Earth's stress, more unrelieved stress may also be building so that earthquakes could occur where there is creep as well as where there is not. We have as yet no measure of the strain accumulating in the "locked" portions, so we have no way of knowing how soon the strain will be too great for Earth to bear, and thus how soon an earthquake will occur.

Our present theory of earthquakes, developed initially from study of the San Andreas, and largely through the evidence of the San Francisco earthquake of 1906, states that stress in the Earth may increase until the rocks of Earth reach a point of strain that they cannot accommodate and break, just as a stick of wood will, when stressed, first bend, then break.

Faulting plays a strong role in the theory of global tectonics, as it plays a major part in the construction of landforms. Indeed, we have not always appreciated the vast alterations that faults have made in our landscape. Most

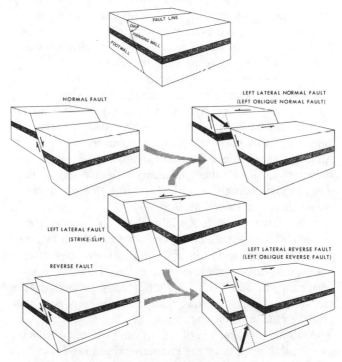

FIG. 39. Types of faults.

people can recognize volcanoes (if they are active or new); most of us can tell when we are in arid country; and everyone is familiar with a river and the seashore; but faults are more difficult to recognize. Faults look very well defined on a geologic map, but identifying them on the ground is difficult.

Except for the volcanic peaks (which may also take their origin from the movement of Earth plates), most of California's mountain ranges are faulted ranges. The vast bulk of the Sierra Nevada is a "fault-block" range; the long mass of the Coast and Transverse ranges is bounded by the San Andreas fault; the east face of the Warner Mountains, in northern California, looking out over the Nevada desert, is outlined by a fault. A glance at a fault map of the state shows the prevalence of faults. There are far more faults than can be shown on a map. Only the Central Valley seems devoid of faults, but even this is deceptive, as there are faults at depth that have been covered by detritus washed down from the surrounding mountains.

One whole region of California, called here the "Basin Ranges" or "Basin-and-Range" province, consists of fault-lifted mountain ranges and intervening valleys. It is part of a much larger area of northwest-trending fault-block mountains known as the Great Basin that covers much of the American Southwest. On some maps, these ranges are shown by a lined pattern ("hachures") that makes them look, as one geologist described it, "like an army of caterpillars marching toward Mexico."

It is relatively easy to see large faults on satellite images or high-altitude aerial photographs, because from that distance the texture and grain of California shows clearly. It is not so easy to recognize even large faults on the ground, where they blend in with other topography. In fact, it is the fault-related features, rather than the fault scarp itself, that give the clue to its presence.

Although the skeleton of a fault can be noted on a good topographic map, or on the images relayed to Earth by satellites in space, to locate each strand requires that even a

FIG. 40. Broken end of an overpass destroyed in the San Fernando earthquake of 1971.

trained scientist work in minute detail. Geologist Dorothy Radbruch-Hall, one of the pioneers of the technique of close observation of faults, has described this phase of her work as "pushing a peanut with my nose."

Scientists have learned, however, that certain features are characteristic of faults. Easiest to see are the changes in man-made structures: houses cracked, curbs and streets torn apart, fences and roads abruptly offset.

It is not as easy to see changes in the ground itself. It is true that furrows in the earth, if associated with an earthquake, may delineate a fault zone. The furrows are not clear evidence in themselves, however, because many other things can cause them. Certain abrupt cliffs (scarps) mark fault zones, and, certainly, a cliff produced after an earthquake where none was before is remarkable evidence of faulting. But, unless a person had been thoroughly familiar with the landscape before the earthquake, he or she might not realize that the cliff was a newly created feature. For example, it is not easy to see the actual line of the fault that produced the San Fernando earthquake of 1971, yet the earthquake was felt by millions of people, triggered thousands of landslides, and lifted the San Gabriel mountains upward some six feet.

California is laced with many fault-line scarps from prehistoric earthquakes—the whole eastern front of the Sierra Nevada is a zone of scarps; but in very few places is it possible to see fault scarps that we know were produced during specific earthquakes. One scarp, still visible near Lone Pine, was produced by the earthquake of 1872. However, you must look carefully, for if you are not sophisticated in recognizing faults, you may not see it!

It is easier to recognize many of the secondary features produced by earthquakes and faults than to identify fault lines themselves. Lakes or "sag ponds" along the fault line are some of the most obvious features; offset streams are another. For example, tiny Diaz Lake in Lone Pine is a relic of the 'quake of 1872; Crystal Springs reservoir, a natural lake enlarged for storage, and lakes north and south of it mark the San Andreas fault near San Francisco. They are earthquake formed, but were not wholly created during the

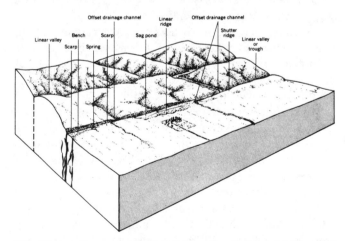

FIG. 41. Landforms developed along active horizontal (strike-slip) faults.

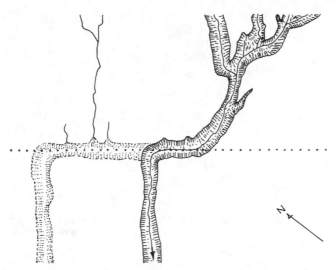

FIG. 42. Movement along the San Andreas fault is shown by these offset stream parts. At one time, the stream may have flowed in a straight line across the fault. Movement along the fault has caused the stream to bend. The stippled area on the left is an abandoned channel that has been stranded by fault movement. The distance from the abandoned channel to the present one is about 400 meters (1,200 feet). Forty earthquakes would be required to produce this much offset with an average movement of about 10 meters (32 feet) each. This sketch was made from photographs of the San Andreas fault where it passes through the Carrizo Plain.

great 'quake of 1906. Farther north, the San Andreas has altered the landscape by moving pieces of land in such a way that Tomales and Bolinas bays have formed in its track. Southward, faulting formed a giant "graben" (German for "grave"), dropping the land between the San Andreas and San Jacinto faults. Later, the flooding Colorado River rushed in to fill the low spot, creating the Salton Sea.

Within a fault zone, old or new, one can sometimes see evidence of the power of past earth movements. Polished rock surfaces with grooves (called by the German word, "slickensides") are one clue. So is finely ground rock flour, which shows that the movement of the two sides of the fault

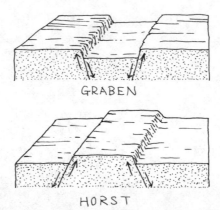

FIG. 43. *Upper.* A *graben* formed by the dropping of a block of land along two faults. *Lower.* A *horst* formed by the uplift of a block of land between two faults.

has had the effect of a giant mill, grinding hard rocks into powder. The powder, called "fault gouge," can be of any color, but much is black, setting it apart from rock on either side of the fault. Along some ancient faults, gouge has been lithified into solid rock. At La Grange hydraulic gold mine near Weaverville the force of water from the huge hoses used in mining has washed away the gravel surrounding a fault zone, leaving the more resistant lithified gouge standing as a low rock ridge in the landscape. An even older fault zone (its gouge now changed to stone) has formed the South Fork Mountain Schist, a 240-kilometer- (150-mile-) long shear zone that separates the Klamath Mountains from the Coast Ranges.

Even after one recognizes a fault, it is not easy to tell whether or not it is active; that is, whether it should be called an "earthquake" fault. All faults that are known to have moved in historic time (that is, have been responsible for earthquakes) are classed as "active." Geologists working in California, where there are many faults, recognize three types of faults: (1) faults that have moved in historic time; (2) those that are known to have moved during or after the Great Ice Age; and (3) other faults. The first two types are "active," while the large group of "other" faults must be given the Scottish verdict, "not proven." It is quite possible that some of them might also become active.

At close view, it is easier to see small breaks called "joints" (because there has been no earth movement along them) than it is to see major faults. In some rocks, joints give a vertical aspect to landforms, as the jointing at Devils Postpile has. These are cooling joints, developed as the rock consolidated from a fluid lava flow.

Joints in granite are more common but less clearly understood. They may result from consolidation, from the removal of pressure, from tectonic stress, or from other, unknown causes. Quarrymen use them to find the best way to excavate building stone; they utilize three breaking directions more or less at right angles to one another.

Jointing in granite has contributed to the creation of some of California's most spectacular landforms. The great domes of the Sierra Nevada stand proudly above the valley because they are less closely jointed, while the splintery peaks of the eastern Sierra Nevada project jaggedly above the desert along the Sierran fault zone, split along joint planes. In some places, slots running through the mountains mark the sites of intensely jointed rock. Where one direction of jointing is not strongly dominant, the rock may weather to "loaves," or may look like a miniature Stonehenge.

When well-jointed rock forms sea cliffs, the joints provide avenues for salt water to penetrate. As a result, jointed rock along the coast tends to weather into steep cliffs with piles of rectangular boulders at their feet.

In the desert, jointed rock often becomes rounded rocks called "tors," and in deserts and foothills alike, joints and parting in slate have given us "tombstone rocks" by weathering and disinterring (see p. 188).

Joints and faults also assist the process of landsliding, help to determine where streams find it easier to run, and provide passageways for the forces of weathering to commence the breakdown of the rocks of the mountains.

Our concern about the connection between earthquakes and faults is a legitimate one, as our personal safety is at stake. Regard for the danger of earthquakes makes us forget their beneficial qualities, however. To those of us who love mountains, the earthquake is a friend, for there are few mountains and valleys in California that have not been created or changed by faulting. We live in earthquake country, and by these earthquakes and faults our spectacular landscape is in the process of transformation. We "live in creation's dawn," as John Muir once said. We are lucky to do so.

TABLE 12. Tectonic Landscapes of California

Feature	Where to see a good example
Anticline	Wheeler Ridge, San Joaquin Valley (Golden State Freeway near Grapevine Grade)
	Buena Vista Hills
	Garnet Hill, Coachella Valley
	Indio Hills
	Desert Hound, Mormon Point, Death Valley National Monument
	Lenwood, San Bernardino County
	Salton Trough
	Calico Hills
	Signal Hill near Long Beach
	Dominguez Hills near Lomita
Dome	Kettlemen Hills near Coalinga
Earthquake walk (self-guiding)	Pt. Reyes National Seashore
	San Juan Bautista
	Los Trancos Open Space Reserve, San Mateo County
Fault	Kern Canyon (last movement 3.5 million years ago)
	Sierra Nevada (east face near Lone Pine)
	Hayward fault, Hayward
	Hanaupah fan, Death Valley National Monument
	Salton Trough, Salton Sea
	Garlock fault, Mojave Desert, especially near Saltdale
	San Jacinto fault, on campus of San Bernardino Valley College
	Imperial fault, Imperial Valley
	Santa Susana fault, near Van Norman Dam, Los Angeles County

TABLE 12. Tectonic Landscapes of California (Contd.)

Feature	Where to see a good example
	West side of Coyote Mountain near Borrego Springs
	(See also San Andreas fault)
Fault line scarp (recent)	Death Valley fault zone, especially Hanaupah Canyon
	Coso Hot Springs, Inyo County
	Lone Pine, Inyo County
	Salton Sea
	Crater Mountain near Bishop
	Calexico
	See also San Andreas fault
Fossil fault zone	La Grange hydraulic mine, Weaverville
Graben	Death Valley
	Salton Sea
	Diaz Lake, Owens Valley
	Clark Valley, Santa Rosa Mountains
	Saline Valley
	Kern Canyon
	Owens Valley
	Tahoe Valley
	Panamint Valley
	Santa Clara Valley
	Hollister Valley
	Wildrose Canyon, near Death Valley
	Garlock fault near Goler
Horst	Panamint Range
Offset fence	Nyland Ranch, 1 kilometer (half a mile) north of San Juan Bautista
	Along State Highway 46 just east of Cholame
	See also Earthquake walk

TABLE 12. Tectonic Landscapes of California (Contd.)

Feature	Where to see a good example
Offset stream	Carrizo Plain
	Near Taft
	Pt. Reyes Peninsula
Sag ponds and fault-controlled valleys	Diaz Lake, Lone Pine
	San Andreas Lake, Crystal Springs Reservoir, San Francisco Peninsula
	Tomales Bay
	Bolinas Lagoon
	Lake Elizabeth
	Lake Hughes
	Lake Temescal, Oakland
	San Andreas fault near Cajon
San Andreas fault	Earthquake walk, Pt. Reyes National Seashore
	San Andreas fault trail, Los Trancos Open Space Reserve, San Mateo County
	Hollister
	San Juan Bautista:
	Fault scarp at Rodeo grounds, east side of Mission Almaden Winery, La Cienega road (National Register of Natural Landmarks plaque)
	Carrizo Plain
	Anza-Borrego Desert State Park
	Bodega Head
	Bolinas Lagoon
	Indio
	Near Palm Springs
	Between Valyermo and Palm Springs
	Peachtree Valley

TABLE 12. Tectonic Landscapes of California (Contd.)

Feature	Where to see a good example
Step faults	Lava Beds National Monument
	Coso Mountains, China Lake Naval Ordnance Test Station
	Volcanic Tableland north of Bishop
Syncline	Palmdale (2 miles southwest on Antelope Valley Freeway)
	Ventura Basin, Topa Topa Mountains
	Pt. San Pedro, San Mateo County
	Moss Beach, San Mateo County
	Painted Canyon near Mecca
	Superstition Hills
	Santa Rosa Mountains
	Rainbow Basin, north of Barstow

8 · WHERE THE LAND MEETS THE SEA

Nowhere is the rhythm of Earth more apparent than at the seashore. Living things have their rhythms, as we all know; great rivers have their rhythms; volcanoes have their rhythms; deserts have their rhythms; probably faults have their rhythms too, if we could but discern them. Some of these rhythms are taken largely from the rhythm of the sun and moon as is the rhythm of the sea; perhaps others are, as well, but we do not know how.

The rhythm of the sea is so hypnotic that one tends to believe it is a constantly repeated pattern. And so it is, but only in a very large sense: waves break on shore and retreat minute by minute; tides rise and fall by the day; great tides rise in the spring and fall. But the pattern differs from day to day and from hour to hour, and each wave within the pattern differs from the last; each tide differs from the one before; and great tides are not the same from year to year.

I learned this quickly and practically once when I was making a time-lapse film of beach processes. Each shot was taken on a different day at slack water. Because the time of slack water varies from day to day, the Sun time of each shot was different. The shots were all from the same vantage point, so that they encompassed the same scene. I wanted to move smoothly from shot to shot, so I decided to try to cut—that is, to change shots—on the crest of a wave at the same place in each picture frame.

The result was disaster. No wave was ever at the same

height, at the same angle, or of the same type as the one before.

Waves form as "trains" wherever the water is disturbed. Most ocean waves are wind waves, derived from storms somewhere. The height of a wind wave depends on how strong the wind is, how long it blows, and the length of open water it blows across (the "fetch"). As these factors vary from moment to moment, it is easy to see that wave trains set up by the wind are not uniform. Tides also make waves, as do passing boats, diving ducks, surfacing whales, and myriad other sources. The Earth itself may shake, causing seismic sea waves. Each wave may encounter an obstacle—which may be another wave—causing it to bounce directly back (reflect) or to bounce off at another angle (refract).

Each of these sources generates wave trains of different heights from different directions, all of which go to make up what sailors call a "sea."

Waves have more shape than substance. Because waves have shape, they have dimensions: the high point is called the crest; the low point, the trough; the vertical distance between adjacent crests, the wave length; the time it takes for a wave to travel one wave length is the wave period.

It is the wave form only that moves in the open ocean, not the water particles. Each wave is made up of particles of water orbiting in circles; the diameter of each circle is the wave height. This is easy to prove in the laboratory, where one can see orbiting balls in a wave tank, but on the open sea it is hard to imagine that a mere shape has made one ill!

Waves of many different periods, ranging from ripples with periods of fractions of a second through wind chop with periods of 1 to 4 seconds and swell with periods of 6 to 16 seconds to tides with periods of 24 hours, may all be present in the sea at one time, adding to the confusion of waves.

All of this may be complicated by such sudden, impulsively generated waves as seismic sea waves (also called

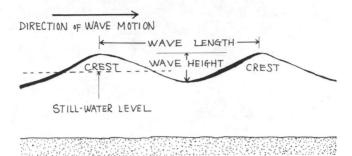

FIG. 44. The parts of a wave.

"tsunamis," or incorrectly, "tidal waves"), by unusually bad storms, by surf beat (a reinforcement of waves by other waves, causing them to increase in height) and by currents, which may themselves have waves.

Waves move toward shore. If the water is deep enough, they may strike the shore directly and be reflected back. If the water along the shore shallows, as it does where beaches are present, waves may break to reform into smaller waves, perhaps break again, reform, and finally rush upon the beach as foaming swash. When a wave moves into shallow water, the wave shape changes from a rounded form to one with a peak; when it reaches the spot where the depth of water is less than half the wave height (that is, where water particles in the waves can no longer make a full orbit), the wave will break. If the wave has broken on an offshore bar, it may reform beyond the bar into another wave and repeat the story. Eventually, the wave reaches a spot where it can no longer form breakers, and the wave rushes—water, this time, as well as form—onto the beach.

The shape of the breakers themselves depends upon weather and underwater topography. On a calm day, in an area where the underwater slope is steep and smooth, plunging breakers may form, perhaps hurling water as much as 15 meters (50 feet) into the air with a loud crash. Surging breakers, which rush onto the beach without break-

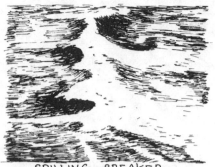

SPILLING BREAKER

PLUNGING BREAKER

SURGING BREAKER

FIG. 45. Types of breakers.

158

ing, also form on steep beaches. If the beach approach is more gently sloping, the breakers may tumble over into spilling breakers of the type favored by surfers.

The line of breakers just beyond the surf zone usually marks the position of an offshore bar, built up by the circulation of sand in the surf zone. Here sand, being pushed along the bottom by the orbiting water in the waves, is dropped by the breakers as they expend part of their energy in breaking. The water in the surf zone, however, moves forward to the shore and then retreats, carrying beach sand with it to add to the bar. As the bar is built higher, smaller and smaller waves will break on it, so that eventually the water over the bar is shallow enough for virtually all waves to break on it. In places, the offshore bar may grow high enough to project above sea level, perhaps creating a lagoon of quieter water behind it.

Bars that form just outside a harbor (by the combined effect of currents that run along shore and ocean waves) may protect the harbor somewhat, but as they also cause the waves to break and "moan" at the harbor entrance, they can endanger craft leaving or entering the harbor. It was this "moaning" that Tennyson referred to when he wrote "May there be no moaning of the bar when I put out to sea."

Not all beaches have protecting bars; on some, deep water continues so close to the shore that the waves crash directly onto the beach (as at Fort Ord, California).

In many places on the California coast, there are at least narrow beaches protecting the shore from the direct impact of the waves. Most California beaches are sand, although a few (such as at Golden Gate National Recreation Area) are made of pebbles; some (as at Point Lobos) are rock; and some are mud (as at Drake's Estero). In winter, beaches are narrower and have more coarse materials; while in summer (when storms are less frequent), beaches become sandier and wider.

Sand to build beaches comes from three sources: the bulk of it is washed down rivers to the sea; part is contributed by

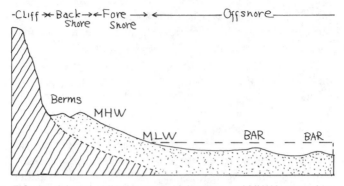

FIG. 46. A beach; MHW is mean high water, and MLW is mean low water.

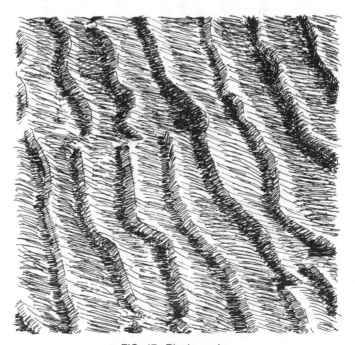

FIG. 47. Ripple marks.

FIG. 48. Shoreline showing beach cusps.

ancient beaches and dunes now covered by the sea; a small part is worn from the rocks near the beach itself. Waves can literally explode on cliffs by trapping pockets of air. These explosions can shove large blocks of rock away from the cliffs into the waves. The sea also uses rocks as tools to hammer at cliffs and as sandpaper to wear away other rock; its salt etches and corrodes firm rock into fragments.

It is during the rush of the surf that the sea grinds rocks to pebbles and pebbles to flour. Here in the surf zone, water tumbles the rocks and sand back and forth as if they were in a giant washing machine. If you stand on the beach, you can hear the machine working: a crash as the waves break, followed by a rushing sound as they approach the shore, then a softer swish as the foam line moves high up on the beach. As the foam reaches its final extent, a soft lisping sound begins; this sound is made by bubbles of air breaking as the water readies itself for a return trip toward the sea. Now the noise of the surf mill can be heard: in beaches where there are large pebbles, it is a rumble and a clatter; on beaches with smaller pebbles, a rattle; on fine, sandy beaches, a shush.

Every movement of the mill rounds and polishes the pebbles, using fine material it has already ground off to shine them. In deeper water, pebbles are less rounded, until, at depths of 10 meters (30 feet) below low tide level, they usually show little sign of wear. Some beaches, where tough rocks are the source of pebbles, have lustrous, polished beach stones. For example, Cronkhite Beach in Golden Gate National Recreation Area near San Francisco has highly polished red and green pebbles, derived from adjacent chert beds. They are beautifully rounded, having been tumbled in the surf mill for about 5,000 years. A few pebbles are torn from nearby cliffs each winter, but most of them were washed from landward rocks during the Great Ice Age, when sea level was lower and the little stream that now drains the canyon behind the beach was much larger.

Cronkhite Beach itself is a pocket beach, one that lies between two projecting headlands. At this particular beach, the headlands are protective enough that currents along the coast do not disturb it much. In fact, the little cove sets up an eddy in the tidal waters, so that the sand this beach holds is washed out to sea in one storm, but is brought back in another.

Most California beaches, however, do not keep their sand. It looks as if they do, but experiments with dye-tracers show that the sand, particularly south of Point Conception, is migrating steadily southward. This movement of sand along the coast is called "littoral drift" (littoral is a word pertaining to shores) and has been compared to a conveyor belt. The sand moves southward and vanishes. Divers, following it, have discovered that the sand being conveyed on this belt is dropped into one of the submarine canyons off the California coast. This belt is comprised of several units, or "cells" in southern California. Each cell commences from a point south of a major submarine canyon and transports sand to the next major canyon, where the sand is dumped.

Under natural conditions, the littoral drift is not threatening the health of the beaches, as rivers from the land bring adequate sand to replenish them; but in California, where almost all rivers are dammed upstream, the supply of sand has greatly diminished. The loss of sand not only robs us of beaches, but the loss of beaches allows the coast to be eroded more quickly.

Besides pocket beaches, there are long, nearly straight stretches of sandy coastline. Some of these are sandy peninsulas built by longshore currents (as at Oceanside and Point Reyes); others are sandy necks called "spits" similar to barrier islands, but attached to the shore (as at Stinson Beach). Sandy strips, spits, and barrier islands are all indications of a tendency of the sea to straighten shorelines and make the "rough places plain." (The word "tombolo" is used for islands connected to the main by a sandbar, as at Morro Rock).

Waves attack bold headlands, tearing rocks from them.

FIG. 49. Morro Rock, connected to land by a narrow ribbon of sand called a "tombolo." The rock is the remnant of a 20-million-year-old volcano.

Because waves strike rough headlands at an angle, they create longshore currents, which carry the torn off pieces across the mouths of quieter bays, building spits and bars as they do so. In this way, both the headlands and coves are protected by the very forces that attacked them.

Some beach sand in California comes from ancient dunes—dunes that were shoreside on beaches of 5,000 or more years ago, before the last rise in sea level that was caused by the melting of glaciers took place. Some of these old dunes may still be seen on shore (as at Dillon Beach and Morro Bay), but many more lie buried offshore under the waters of the Pacific Ocean.

Beach dunes obey the same laws as desert dunes, taking many of the same shapes. Like desert dunes, they form only when the sand is dry, where it is abundant, and where winds are strong and frequent. Dunes once occupied much of the coast of California, but with diminishing sand supply, construction (as on the San Francisco peninsula) and the mining of irreplaceable beach sand (which is still being carried on), it is likely that dunes may eventually vanish from California shores.

Much of the California coast is lined with cliffs. The cliffs range from steep and rugged to rounded and gentle, depending upon the kind of rock of which they are made. Hard, tough, durable rock makes bold, steep headlands, while softer, less resistant rock forms lower cliffs and embayments. The shape and color of coastal cliffs provides much of the scenic interest of the shore. There are striped cliffs (as at Point San Pedro, San Mateo County), light-colored cliffs (as at Point Lobos, Monterey County), dark cliffs (as in Humboldt County), and even cliffs that resemble a colorful cathedral (as at Tennessee Cove, Marin County, where red and green chert beds alternate).

In places, the waves have cut notches into the cliffs to etch out sea caves (as at La Jolla). In horizontally bedded rock, or in jointed rock, the sea may erode completely through a projecting headland to carve a sea arch such as are

FIG. 50. A sea cave.

FIG. 51. A sea arch.

at Natural Bridges State Park, Santa Cruz. An arch may collapse as one in the park did in 1980, or the sea may isolate a pillar of rock to become a "sea stack." Sea stacks are common along the northern California coast (one that is usually covered with birds lies just offshore from Cliff House, San Francisco).

At many places on the California coast, flat, level land lies adjacent to, but above, sea level. Called "marine terraces," these are former beaches cut when the sea was at a higher level with respect to the land. As the coastal ranges in California are very young mountains, just now in the process of growing, the ancient beaches partly reflect the building of the mountains, and partly the melting of the great glaciers. In places, many rises and falls of sea level have left their mark; in some places, fourteen different levels can be counted. The coasting traveler (by sea or land)

FIG. 52. Terraces along the California coast. In the Palos Verdes Hills, 13 terraces have been cut by the ocean in past years, each marking a former shoreline. The highest terrace is 400 meters (1,300 feet) in elevation, indicating that the Palos Verdes Hills have been lifted that much since this terrace was lapped by the sea.

can spot these terraces as building sites, pasture lands, and as spots for lighthouses.

At Pacific Palisades, where travelers along the Pacific Coast Highway can view many marine terraces, fossil evidence indicates that the main old terrace, which stands about 75 meters (225 feet) above present sea level, is about 125,000 years old. If so, the rate of uplift of the land with respect to the sea has been about half a meter (a foot and a half) per thousand years.

Other terraces, including the most recent predecessors of today's beaches—those that preceded the rise of sea level about 5,000 years ago—lie buried beneath the waters of the Pacific. Tidal gauge measurements indicate that the sea is still rising at the rate of about 1 mm per year, or 4 inches per century.

Many rivers emptying into the Pacific along the California shore plunge so swiftly into the sea that deltas have not yet had time to build. This is due, in part, to the last rise in sea level, which covered previous deltas, forcing rivers (which have much less water now) to create new deltas farther inland.

Much of our coast shows clear evidence of this last sea level rise. River valleys have been flooded, creating estuaries; coastal valleys have been transformed into bays, lakes, and lagoons. Overtopping of the land by the sea in this fashion is called "drowning"; one can speak of the mouth of the Pajaro River at Moss Landing as having been "drowned." Drowned, but perhaps not for long. The next mountain-building movement may lift the land above the sea again, raising drowned beaches and making new terraces.

The sea beats relentlessly upon the shore. Rivers move endlessly to the sea. Beaches are built and then eroded; cliffs are torn apart and renewed. Unlike much of geologic change, the transformation of the coast goes on at a fast enough pace that we can see the Earth change while we watch.

An inventory of the California coastline made in the 1970s showed the shoreline configuration to be 11.6 percent tidal lagoons and 14.9 percent tidal estuaries (making a total of 26.5 percent of the coast in bays); 15.1 percent offshore islands; 22.6 percent straight beaches, 28.8 percent rocky coasts, 4.0 percent hooked bays, and 3.0 percent large embayments (making a total of 58.4 percent mainland). But because the shore is constantly changing, its shape today is not what it was when this inventory was taken, nor what it will be in another decade.

Constant change is the nature of the coast, but because we humans want to use the beach, too much change disrupts our plans. Along many of the coastal cliffs, landsliding is common and natural, as at Point Reyes National Seashore. In other places, however, where landslides threaten buildings and other human structures, we press for landslide protection—even though it was our own lack of wisdom that placed buildings in such perilous positions.

The ordinary processes of coastal erosion pose threats to construction at many places on the California coast. Because protecting beaches have been lost, or for other reasons, erosion is acting more rapidly in some places than it might have if humans had not interfered. An example of this type of interference is to be seen in England. The coarse beach "shingle" (rock) was removed from the beach at Hallsands for use in construction work. As the beach rocks protected the cliff on which the town was built, their removal allowed the sea to erode the cliff so rapidly that the entire village of Hallsands was destroyed.

Accelerated erosion of this sort is a problem to several communities in the San Francisco Bay area, including Bolinas, Alameda, Pacifica, Rio del Mar, and El Granada, as well as such areas in southern California as Sunset Cliffs, Blacks Beach, and Newport Beach. The rate can be quite rapid. At Goleta, Santa Barbara County, cliffs are receding at the rate of half a foot per year.

Various protecting devices have been used to control accelerating erosion, such as groins, sea walls, breakwaters. For example, the city of Santa Barbara, California, which had no sheltered port, wished to construct a quiet anchorage. Three proposals were prepared over the years, and three times agencies of the federal government advised against building a permanent breakwater. In spite of this advice, a breakwater, an offshore protective wall open at both ends, was constructed. This design allowed sand to accumulate on the beach and proved to be an inadequate protection against the wind. To reroute the sand and add more wave and wind protection, the western end was extended to shore, thus providing a calm anchorage for a few years.

During that time, sand accumulated alongside the breakwater and finally inside the harbor. By forcing the longshore current to drop its load of sand in the new Santa Barbara harbor, the breakwater had allowed the current to move east and south without its usual load. Thus freed, it was able to accumulate a new load, which it did by eroding the beaches for 16 kilometers (10 miles) east of Santa Barbara. In one stormy winter, as much as 50 meters (150 feet) were cut away from some beaches.

Meanwhile, Santa Barbara's new harbor was filling up. To prevent the building of a beach where the town had intended a harbor, the sand was dredged from the harbor and piled offshore, where it was hoped that the longshore currents would pick it up rather than raid the adjacent beaches. Unfortunately, the placement of the dredge spoil was wrong, and the ridge remains today.

This debacle provided the impetus for modern Pacific coastal studies, which showed that the currents at Santa Barbara moved an average of 585 cubic meters (770 cubic yards) of sand past a certain point on the breakwater daily. In stormy weather, the volume can rise to 3,190 cubic meters (4,200 cubic yards); in summer it may drop to 230–300

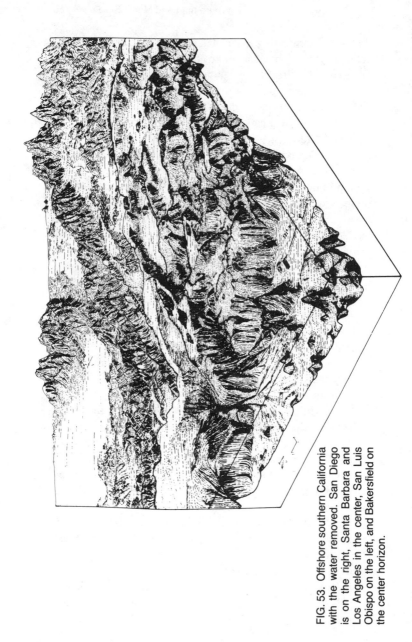

FIG. 53. Offshore southern California with the water removed. San Diego is on the right, Santa Barbara and Los Angeles in the center, San Luis Obispo on the left, and Bakersfield on the center horizon.

cubic meters (300–400 cubic yards). This is not particularly high; elsewhere on the southern California coast the volume is as much as three times that amount.

Our shores need protection partly because, at present, Earth's ocean basins are overfull. If one compares them to an old-fashioned wash basin with a lip, then today's seas are filling the basin so full that water laps onto the lip.

The sea basins are not smooth, like a wash basin, however, but are more rugged than the continents, with deeper valleys and taller mountains. Between the ocean basin and its rim there is a steeply sloping rise forming the edge of the basin, called the "continental slope"; landward is the rim itself, known as the "continental shelf." The controversial offshore drilling for petroleum is taking place on the continental shelf.

Many of us will never see that part of California that lies beneath the sea. Indeed, even those who dive can see only a few feet of the sea bottom at a time; nowhere can we obtain the wide vistas we cherish so much on land. It remains for scientists and sailors to decipher this region for us and for artists to construct a visual rendering of their results.

The northern part of offshore California is a narrow strip of continental shelf paralleling the shore except at San Francisco, where it widens to 48 kilometers (30 miles) to encompass the tiny Farallon Islands, and north of Cape Mendocino, where the Mendocino escarpment forms a bold subsea barrier.

At San Francisco, the sea is unusually shallow just west of the Golden Gate, where a lunate sand bar, deposited by rivers heading in the California mountains and augmented by debris from the days of hydraulic mining, lies just below the water's surface. This bar, called the "Potato Patch," has upset the equilibrium of many a would-be small boat sailor, and, in times of storm, has caused many a large vessel to go aground.

All of us recognize that there are sand bars in the shallow parts of the ocean that are a hazard to sailing, but, until

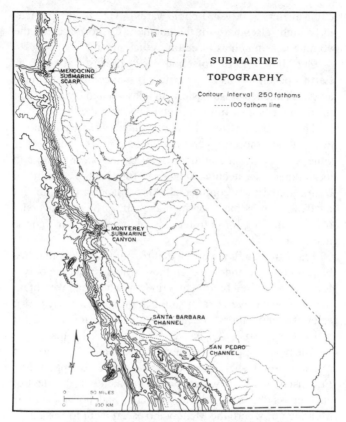

SUBMARINE TOPOGRAPHY

Contour interval 250 fathoms

----- 100 fathom line

MENDOCINO SUBMARINE SCARP

MONTEREY SUBMARINE CANYON

SANTA BARBARA CHANNEL

SAN PEDRO CHANNEL

N

0 — 50 MILES

0 — 100 KM

MAP 11. Submarine topography off California.

recently, few of us had an appreciation of the rugged topography of the ocean basin. We tended to visualize it as
a flat, level plain, sloping gently away from the continents.
Such is far from true. For example, a series of submarine
canyons lies off the California shore that vies with any canyon on shore. At least thirty-two of them lie off southern
California, of which nineteen have names; many others are
adjacent to the northern California shore.

Largest and grandest of these is Monterey Submarine

Canyon, which presently heads in Monterey Bay, but is invisible beneath the blue water. Its width is about 20 kilometers (13 miles), and its depth about 1.5 kilometers (1 mile), making it about as wide and deep as the Grand Canyon of the Colorado.

In the rich petroleum-producing region off southern California, the topography buried beneath the sea is an extension of the hills and valleys of the land. In fact, if sea level were lowered by 100 meters (300 feet), as it was during the Great Ice Age, there would be one island off Santa Barbara where now there are four; San Nicolas would be quadrupled in size, and several new islands would appear from beneath the waters.

Undersea banks, ridges, sea mounts and basins take the place of landward mountains and valleys. At some time in

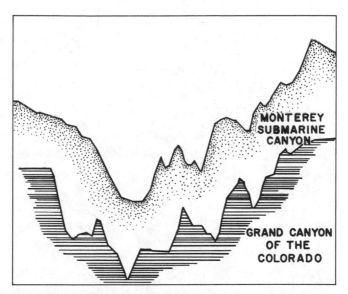

FIG. 54. A comparison of Monterey Submarine Canyon and the Grand Canyon of the Colorado. Both canyons are about 20 kilometers (13 miles) wide and 1.5 kilometers (1 mile) deep.

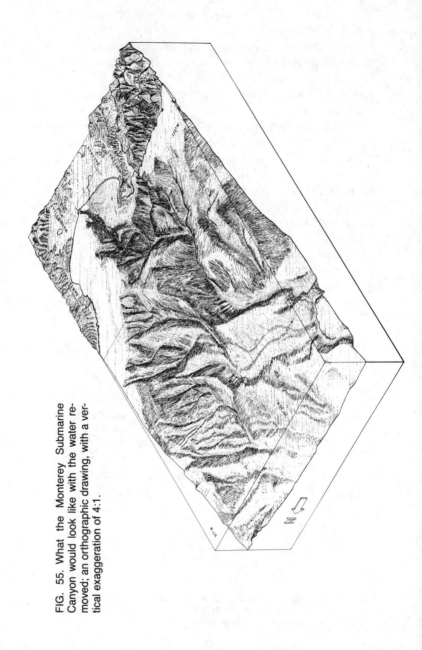

FIG. 55. What the Monterey Submarine Canyon would look like with the water removed; an orthographic drawing, with a vertical exaggeration of 4:1.

the fairly recent geologic past, when the islands were probably connected, a species of dwarf elephant lived on Santa Rosa Island. They are fossils now, but how and when the elephants arrived on the island remain a mystery. Even if sea level were lowered by twice the amount generally accepted as likely during the Ice Age, the animals would still have had to swim several miles if they arrived from the mainland, as some scientists think happened. The fact that they were dwarf elephants indicates that they were in residence for some time; long enough, at least, to have been isolated from other elephants and evolve into a smaller species.

Where the elephants came from and how they got on the island are but two of the many mysteries the sea still holds. We have as yet only the dimmest understanding of what California is like beneath Pacific waters, or what is happening where land meets sea.

TABLE 13. Landforms of the Coast

Feature	Where to see a good example
Barrier beach	Ft. Cronkhite, Golden Gate National Recreation Area
Bay mouth bar	Bolinas Lagoon, Marin County
Drowned river valley	Drake's Estero, Pt. Reyes National Seashore
	Bolinas Lagoon
	San Francisco Bay
	Lagoons in San Diego County, including Batequitos, Buena Vista, and Aqua Helionda
Estuary	(See Lagoon)
Fault line	Bolinas Bay, Marin County
	Tomales Bay, Marin County
	Bodega Head, Sonoma County
Lagoon and estuary	Dry Lagoon Beach State Park
	Big Lagoon, Freshwater Lagoon, and Stone Lagoon, Humboldt County
	Humboldt Bay
	Drake's Estero, Marin County
	Buena Vista Lagoon, San Diego
	Mugu Lagoon, Pt. Mugu
	San Elijo Lagoon, San Diego County
	Batequitos Lagoon, San Diego County
Landslide	Pacific Palisades
	Pt. Reyes National Seashore
	Palos Verdes Peninsula
Marsh	Suisun Bay, Solano County
	Mugu Lagoon
	Sunset Beach
	Pt. Reyes National Seashore

TABLE 13. Landforms of the Coast (Contd.)

Feature	Where to see a good example
	Carpinteria
	Up and down the California coast
Sand dune (coastal)	(See map 17)
Sand spit	Stinson Beach, Marin County
	Humboldt Bay, Humboldt County
	Morro Bay, San Luis Obispo County
	Mission Bay, San Diego County
	Goleta Beach, Santa Barbara County
	San Diego Bay from Imperial Beach to Coronado
Sea arch (or bridge)	Natural Bridges State Park, Santa Cruz
	Anacapa Island
	Pt. Reyes National Seashore
Sea cave	La Jolla, San Diego County
	Kirby Cove, Marin County
	Ocean Beach, San Diego County
	Palos Verdes Peninsula
	Monterey shore
	San Luis Obispo shore
	Mendocino–Sonoma Coast
	Pismo Beach
	Pt. Reyes National Seashore
	Santa Cruz Beach
Sea cliffs	Davenport, Santa Cruz County
	Pt. Reyes National Seashore
	San Mateo beaches (especially Pt. San Pedro and Devils Slide)
	Santa Cruz
	Santa Cruz

TABLE 13. Landforms of the Coast (Contd.)

Feature	Where to see a good example
	Humboldt, Mendocino, San Luis Obispo, and Monterey County beaches
	Northern California coast
Sea stack	Trinidad Beach, Redwood National Park
	Pt. Reyes National Seashore
	Pt. Buchon, San Luis Obispo County
	Up and down the California coast, especially north of San Francisco, along Monterey and San Luis Obispo County coasts, and near Eureka
Submarine canyon	(See figures 53, 54, and 55)
Terrace (marine)	South of Carmel
	San Diego
	Mendocino City
	Ft. Bragg
	Pt. Vicente, San Pedro Hills (Palos Verdes)
	Encinitas
	Palos Verdes Hills
	Halfmoon Bay, San Mateo County
	San Clemente Island
	La Jolla, San Diego County
	Dana Point, San Diego County
	Montara Point, San Mateo County
	Moss Beach, San Mateo County
	Duxbury Reef, Marin County
	Santa Cruz
	Ventura
	Bolinas, Marin County
	San Clemente, Orange County
	San Onofre Mountain, Orange County

TABLE 13. Landforms of the Coast (Contd.)

Feature	Where to see a good example
	Corona Del Mar, Orange County
	Santa Barbara
	Pt. Buchon
Tombolo	Morro Rock, San Luis Obispo County
	Big Sur, Monterey County
Underwater pinnacles and cliffs	Whittier Reef, San Diego County
	San Clemente Island
	Morro Rock, San Luis Obispo County
Underwater reef	Whittier, San Diego County
	Duxbury, Marin County
	(Not organically produced)

9 • THE DYNAMIC DESERT

When one thinks of the desert, sand dunes and waterless lakes, rivers of sand, naked volcanoes, and stark mountains in a rugged landscape come immediately to mind; but what truly separates an arid land from others is its dry climate and sparse vegetation.

The forces of change in the desert are wind and water, as they are in other climates of Earth. Yet because of the dryness, the face of the desert is different. Deserts can be hot or cold, high or low, near the poles or near the equator. An arid belt lies on each side of the equator between 30° and 50° latitude where trade winds blow over hot interior lands. In these belts lie the deserts of northern Mexico, North Africa, Arabia, central Asia, China, and South America.

Along the west coast of South America and southern Africa, where cold ocean currents meet the hot land, there are strange deserts beside the sea that are shrouded in fog, but where almost no rain falls. The Atacama Desert, in southern Peru and Chile, is one; it has spots that receive less than 0.1 centimeter (0.04 inch) of rain per year.

There are deserts on the tops of some high mountains, where rain falls below but does not reach to the crest. Some of these deserts are cold, by virtue of their elevation, but the largest area of cold arid lands surrounds the two poles of Earth.

Most of the deserts of North America have been robbed of water by mountains that intercept rain before it can reach the thirsty deserts. In this manner, lands of eastern California and Nevada are left dry because they lie in the rain shadow of the high Sierra Nevada.

North American deserts are unexpectedly rugged. They consist, not of lands flat to the horizon, unbroken save for rippling dunes, but of mountain ranges, high and low, alternating with flat-floored valleys.

Ranges in the California deserts are not so regularly aligned north–south as those farther east, but instead are oriented in various directions in response to faulting and other Earth forces. Here, the controlling faults are not dominantly north–south; a glance at map 10 will show that the master San Andreas curves eastward near its junction with the east–trending Garlock.

Although the western deserts lie largely in the Great Basin province, many of them have individual names. In California, for example, the Mojave and the Colorado are distinguished as separate deserts, although both belong to a larger unit, the Sonoran Desert. They, as well as Death Valley, the Anza-Borrego Desert, and others, are part of the drylands of the southwest, and share the character that sets the desert apart from other regions.

Not all of the Great Basin is true desert, as judged either by its rainfall or its vegetation. Although the criteria that distinguish a desert are not agreed upon, all researchers agree that deserts are dry. Just how dry is the question. By an old definition, arid lands cannot average more than 25 cm (10 inches) of moisture each year. Another, more modern line is drawn at 13 cm (5 inches) of moisture each year. The Mojave and the Colorado deserts and Death Valley easily qualify by this definition, but much of eastern California is borderline. Rainfall at Owens Lake, for example, is about 15 cm (6 inches), while rainfall a few miles east in much lower Death Valley averages only 3 cm (1½ inches) per year. Another definition requires that evaporation and use of water by plants exceed precipitation. By this definition, areas near the San Diego coast are desert. At Yuma,

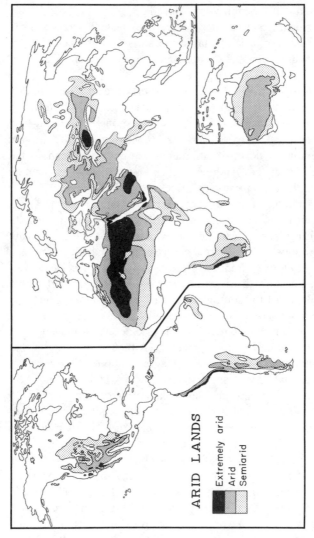

MAP 12. The world's deserts.

ARID LANDS

Extremely arid
Arid
Semiarid

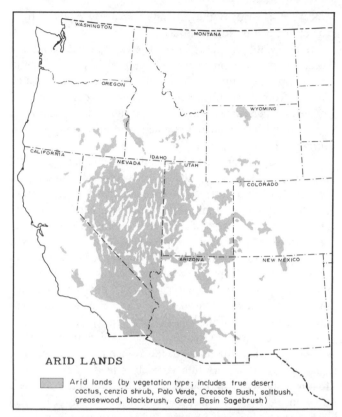

ARID LANDS

Arid lands (by vegetation type; includes true desert cactus, cenzia shrub, Palo Verde, Creosote Bush, saltbush, greasewood, blackbrush, Great Basin Sagebrush)

MAP 13. Arid lands of the American West.

Arizona, near the California border, evaporation and transpiration together are 35 times the rate of precipitation.

Defining deserts so mathematically may be useful to a scientist, but the casual traveler has little opportunity to gauge rainfall or measure the life processes of plants. Here is a rule-of-thumb definition that may not be as rigorous, but is much quicker: if there are no vascular plants (those with roots, stems, and leaves), the spot is extremely arid; if vascular plants are spotty, growing only in the most favorable places, it is arid; if the ground is sparsely covered by a variety of plants, it is semi-arid.

Because they support little vegetation, desert soils are lean, never developing into the deep, rich soils of humid and temperate climates. Rich soil requires humus, which derives from decaying vegetation. What vegetation there is in deserts desiccates when it dies, rather than decaying. This provides a good environment for preserving fossils, but a poor one for building soil.

Little vegetation yields lean soil; lean soil supports little vegetation. It seems as if the desert were always fated to be desert. Yet the desert, when watered, can be very fertile. The fertility comes from the release of potassium from minerals in the thin desert soil. In wetter regions, where soil has been forming for millennia, potassium was long ago used by the plants that made the soil. Farmers need to be cautious in watering the desert, however, as it is easy to pull salts in the soil upward by capillary action. If salts do rise, they may form an alkali "hardpan" or caliche layer that makes the land unfit for further agriculture. Caliche is now forming in the California Great Valley, as well as in the desert.

Partly because there are so few transpiring plants to produce haze, the air is often very clear in the desert. Earth features stand out stark and sharp in the desert air. At least, they once did; California's desert dust storms (some of them generated by human activities) and creeping smog are limiting our view.

Although most of the Earth's cold deserts surround the two poles, California has one located in the White Mountains along the Nevada border. Here, where elevations reach 4,300 meters (14,000 feet), the land is robbed of water by the bulwark of the Sierra Nevada to the east. Winters are cold and dry. Odd landforms, called stone stripes and stone rings, the pebbles in them arranged by frost action, are to be seen on the mountain slopes.

The remainder of California's deserts are hot deserts. Although the world's heat record is held by the Sahara Desert, where 57°C (136°F) was recorded, Death Valley is not far behind, having registered a searing 56.5°C (134°F). Ground temperatures are much hotter, reaching as much as 88°C (190°F).

Desert heat is not steady heat. It, like the wind, rises in the day as the sun climbs, but may vanish at night. Fifty degrees (F) of temperature difference between day and night are not uncommon especially in winter. As writer Ruth Kirk has observed, this means that the lover who swore to be true " 'til the sands of the desert grow cold" need only wait until about three hours after midnight.

One striking characteristic of the desert is that its plants, sparse though they may be, are not those of more humid regions. They are especially adapted to the dry weather by a variety of means, all of which give them a different aspect.

If one were to follow plants that mark a watercourse (which may be dry), one would discover another unusual feature of deserts: water does not necessarily find its way to the sea. Rivers sink into the sand, or empty into lakes that become dry or that have no outlet. "Interior drainage" is thus a hallmark of deserts, flouted only in rare instances by streams that start in more humid regions, flow through deserts, and eventually end in the sea.

Deserts differ from other regions of the world in that they are dry, yet most of their landscape features are not products of the arid climate. Some are independent of climate; some are inherited from earlier, wetter climates.

In some deserts, almost the only visible water is running in rivers that are merely passing through. Such rivers are life-giving, creating ribbons of oases along their banks. The Nile, the Niger, the Tigris-Euphrates, the Amu Darya (Oxus), the Colorado—all of these rivers carry water through deserts from snowy mountains to the sea. Unlike other desert streams, they have a supply outside the desert sufficient to prevent them from drying up; in addition, they have an outlet beyond the desert.

Volcanoes and faults, too, have not come into being be-

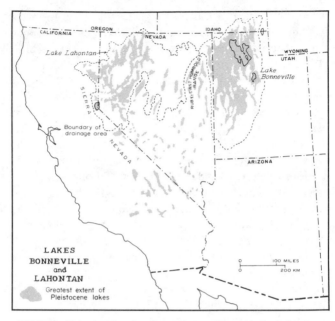

MAP 14. Extent of Ice Age Lake Bonneville and Lake Lahontan.

cause of the desert climate. Quite the reverse may be true; faults and volcanoes may help to create the desert by altering topography. Volcanoes and faults seem more dominant in deserts than elsewhere because they can be seen better; there is no vegetation to mask or soften the new contours, and water is too scarce to attack them continuously, as it does in wetter climates. They stand out sharp and clear— cones, domes, rumpled lava flows, spires, ash falls.

Faults, also, are more easily seen. Faults bound many of the desert mountain ranges: faults, whose movement lifted the Sierra, made desert of the land beyond; faults that lifted the mountains flanking the Salton Sea left the Colorado Desert in a low, now arid basin; another set of faults dropped Death Valley—or lifted the mountains around it.

There are other Earth features, too, that just happen to

be in deserts. Hot springs, which are related to volcanoes and faults, occur in the desert as well as along the seacoast or along river courses. Craters may be created by the impact of meteorites ("falling stars") wherever they hit Earth. Like volcanic cones and faults, however, they are best seen in the desert where vegetation does not conceal and where erosion proceeds slowly.

The California desert is dry today, but it has not always been so. During the Great Ice Age, the climate was often wetter than now. Snow dropped in the mountains, forming glaciers, while rain and snow fell on the land at the foot of the mountains and in today's desert land. One reason the desert was wetter is that the Sierra Nevada and coastal mountains were not as high as they are now, so that winds could blow moisture-carrying clouds from the ocean over their crests to the land beyond.

During the latter part of the Great Ice Age, two vast lakes covered much of the desert West. Lake Bonneville occupied much of Utah, while Lake Lahontan lay in Nevada, with extensions into Oregon and California. Besides these two large lakes, there were many small ones scattered through the West. One chain, which led from Lake Mono to Death Valley, must have looked like a string of blue beads from the air. It can still be seen, but most of the "beads" have become yellow and sere.

A few have not: Mono Lake, for one, is not yet gone, although it is shrinking rapidly. Sixty thousand years ago it may have been connected to the north with ancient Lake Lahontan and it may have spilled east through Adobe Valley. At any rate, it did spill over to the south into Long Valley, where a large natural lake then existed. This lake, in turn, fed the ancestral Owens River, which flowed into ancestral Lake Owens. Then, Lake Owens was, as it continued to be until robbed of its water in this century, "hauntingly beautiful." (Ice Age lakes are called, for example, "Lake Owens" to distinguish them from modern counterparts, which geologists try to name in reverse, as "Owens Lake.") To-

FIG. 56. Tors of the desert. Most desert travelers and movie buffs are familiar with these rounded rocks that stand in desert country. Many of these piles look like misshapen clusters of balls (one group in Australia is called "The Devil's Marbles"). In the United States, we are familiar with them as components of scenery in Westerns; in fact, so many films have been made in the groups of tors known as the Alabama Hills, near Lone Pine, that a scenic road through them leads to "Movie Flat" and other filming locations. In Africa, piles of rocks like this are the lair of lions.

Although we commonly see them in deserts, they take their origin from wetter days of the Great Ice Age (Pleistocene Epoch) that ended about 10,000 years ago. During much of this time, California's tors were buried under wet ground. They were not yet tors; rather, they were large masses of granite containing incipient cracks or joints. As the masses weathered, they broke into squarish blocks. When the soil covering and surrounding them was washed away in ensuing years (in the case of the Alabama Hills it filled in Owens Valley), the edges of the square blocks were rounded, leaving piles of roundish granite knobs. The tors in this sketch are in Joshua Tree National Monument, California.

day, Mono Lake has sunk too low to use its natural spillway to the south; thus no water reaches Owens River from there. In 1913, completion of the Los Angeles aqueduct began to divert most of the water in the Sierra Nevada that once reached Owens River, so that source, too, is gone. Gradually the lake has dried, leaving a brown skeleton—a playa—where once waves washed to shore and steamers docked.

The course of the Owens River was changed at least twice on its way from Lake Owens to the next lake in the chain. Lava flows, pouring from the cones and domes that still dominate the landscape at the foot of the Sierra Nevada south of Owens Lake—the Coso country—pushed the river to one side. The river responded by pouring over the cooled lava in waterfalls that are still visible, though no water plunges over them. At "Fossil Falls," near Little Lake, one can see potholes where the water swirled, the stones that ground them still lying within.

Beyond Fossil Falls, the river dropped into lakes in Indian Wells Valley, China Lake Valley, Searles Lake Valley, Panamint Valley, and finally (at high water stage) Death Valley. Today all these lakes are gone. Of the entire chain, only Mono Lake remains as a relic of bygone splendor. The rest have become playas—dry lakes that only in the wettest of times have water in them.

Mono Lake is one of the few pluvial lakes anywhere in the West that is still a year-round lake today. It is about 23 kilometers (14 miles) long by 16 kilometers (10 miles) wide, and its greatest measured depth of water in this century was 51.5 meters (169 feet). Although the lake and its basin have been modified by fairly recent volcanic activity (an eruption may have occurred underwater as recently as 1890), Mono Lake is one of the few places in the world where the relationship between pluvial lake height and glacial stages is quite clear. Mono Lake's Pleistocene predecessor, Lake Russell (named for I. C. Russell, who worked out the geologic history of the lake) was about 300 meters (900 feet) deep. It left its mark at various lake levels as terraces cut into the hills, giant "bathtub rings" in the desert. In places, the terraces are cut into glacial moraines, showing very clearly the relative age of each.

The water of Mono Lake is very alkaline—so much so that swimming in it is unpleasant and drinking it is dangerous. At present, the level of the lake is falling rapidly, as its source waters, which come from the Sierra Nevada, have

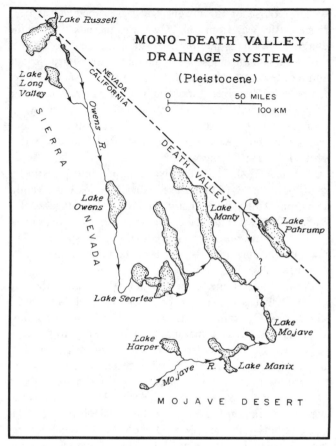

MAP 15. Drainage systems of southeastern California during Ice Age time.

been diverted for use by the city of Los Angeles. As the lake level falls, the waters increase in salinity; someday—soon, perhaps—it, too, will be a playa.

The two picturesque islands in Mono Lake have different origins. Black, brooding Negit is a volcano; light-colored Paoha is sedimentary, with hot springs from which it takes its name: "Paoha," the Mono Indian word for "spirits."

At some playa lakes and at relict lakes Mono, Honey,
and Pyramid, curious white and brown towers called "tufa
domes" mark the shore. One researcher has counted 12 different types of tufa from old Lake Lahontan. Tufa is merely
limestone—calcium carbonate—but takes forms reminiscent of ruined towers, mandalas, and melted honeycombs.
Tufa domes at Mono Lake are still forming, perhaps by

MAP 16. Playas of California.

virtue of springs in and near the lake. Tufa towers at Searles Lake are 30 meters (100 feet) high, giving one an idea where the lake stood when full. They are not now growing. Some of the tufa was probably formed by algae, some by purely inorganic action.

More than 50 playas can be counted in southern California alone. Many of them have had long and eventful histories, as one can read by examining the layers of sediment stacked in them.

At Searles Lake, for example, layer upon layer of sand, silt, and salt has been piled to a thickness of several hundred meters over a 10,000-hectare (25,000-acre) lake bed. The layers are now mined for their content of borax and various other salts. At China Lake, another in the Death Valley–Mono Lake system, a twig buried under 2 meters (7 feet) of sediment proved to be 3,500 years old; this gives a rate of deposition of about 1 meter in 1,750 years, or a foot in 500 years.

Where did this blanket of salts and silt come from? It came from the mountains of the desert—from rocks as the sparse ground water dissolved soluble minerals; from rocks as the occasional desert flood carried broken fragments;

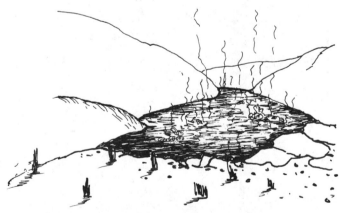

FIG. 57. Hot pool in the desert, fed by hot springs.

from rocks as they weathered to form clay which then floated to a new resting place. By these normal earth processes the dry lake's skeleton was built.

One may wonder if the rocks of which the mountains are made have enough of the constituents of the blanket to account for as thick a layer of salts as is in the valleys. Again, the answer is yes; Charles B. Hunt, who has studied Death Valley intensively, has calculated that the rate of erosion and the chemistry of the rock of the desert mountains that are Death Valley's watershed are adequate to account for the salts in Death Valley. The desert's system of internal drainage, which keeps water trapped, simply keeps the salts in the desert rather than adding them to the sea.

Death Valley is a good example, too, of how a desert lake dries. About 5,000 years ago, antique Lake Manly stood in the heart of Death Valley. It was only about 10 meters (30 feet) deep—not deep for an Ice Age lake (Lake Bonneville and Lake Russell were over 300 meters (900 feet) deep at one time, and Lake Lahontan was over 200 meters (700 feet deep). As the years passed and the climate grew hotter and drier, Lake Manly gradually evaporated. The salts in it crystallized, one by one. The least soluble salt (that is, the one that dissolves with the most difficulty) was the first to drop out of solution. It crystallized on the outer edges and on the bottom of the lake; in a ring next to it on the outer edge and above it in a layer on the bottom, came the next least soluble salt, and so on, until the topmost layer in the center of the basin is sodium chloride—table salt, the most soluble of all. Altogether, nearly 40 different salts rim Death Valley's salt pan, including various forms of sodium and calcium salts, borax minerals, and sulfur.

Death Valley's salt pan (the center of the valley) has a rough and rumply surface. It is a miniature mountain range, with individual peaks about half a meter (two feet) high. From a distance it looks smooth; close up, it obviously is an intricate salt badlands that has been etched by rain and trimmed by wind.

Not all playas have rough salt surfaces. Some, like Bonneville playa in Utah, have salt crusts so smooth they can be used as speedways. Some, like Racetrack near Death Valley, have smooth clay crusts which, depending on the level of ground water, can be wet or dry. These clay playas are the flattest natural surfaces on earth. One playa, Rogers Dry Lake, is so hard and strong that the space shuttle, *Columbia*, landed there at the end of its voyages; some are so soft that an unwary walker can be mired. Some have a puffy surface called "self-rising ground," where ground water has been lifted by capillary action and caused the clay to expand into large puffs.

Capillary action also helps to form an alkali crust called caliche in the American Southwest. It is one of several forms of hard crust ("duricrust") that develop when various chemicals in ground water are precipitated in the soil. Caliche (also called "hardpan") is calcium carbonate—limestone. Other forms of crust are ironstone, an iron-rich variety, silicrete, a silica-rich variety, and a form rich in sulfate. Silica-rich and iron-rich caliche are usually formed in more tropical climates, but they can be seen as brown bands in desert rocks—relics of warmer times.

Caliche forms when ground water is pulled upward by capillary action, and then evaporates. Where the water evaporates, minerals in solution are precipitated out in a layer. Or a layer of caliche may be formed by overwatering the desert soil. By flushing the minerals in the soil with water, they are dissolved and forced downward. Caliche can form in these conditions by downward movement of water, as well as upward, capillary migration. If for any reason the soil is removed (such as by a flood or by wind after plowing when the ground is dry), only a layer of hard alkali crust will remain.

Desert playas may be rough or smooth, salty or clayey, soft or hard, but almost all of them have mud cracks. Cracking takes place after a playa has been flooded, which allows the clay to swell. As the swollen clay dries, it shrinks, forming cracks that outline more or less regular polygons. If the clay dries evenly, straight cracks meeting at 180° will form; if it dries unevenly, curled cracks meeting at 90° are more likely. If conditions are ideal, a six-sided pattern like a honeycomb will develop.

Most mud-crack polygons are a third of a meter (a foot) or less in diameter, although there are giant polygons on at least 30 playas in the western United States that are visible only from the air or from high places. Some are as much as 100 meters (300 feet) in diameter. Intermediate sizes have not been seen in the North American deserts.

If the center of a mud polygon is slow in drying, the edges may curl upward. Along these edges, or in very thin clay layers on playas that are lightly wetted, tiny, fragile mud curls may form that are light enough to be blown away by the wind.

Cracks surrounding the mud polygons may be narrow or wide, shallow or deep, depending upon the type of mud and the length of drying time. If the mud contains a great deal of swelling clay, the cracks will be wider, while the depth of cracking is determined by the speed and thoroughness of drying. More clay in the playa surface causes the cracks to be closer together; more sand encourages wider spacing. As one might guess, clay cracks that have been drying a long time are the widest and deepest.

Sometimes wild desert storms will fill the cracks and cover a dry lake with a layer of sand. When the lake is next flooded, clay of the polygons may swell, pushing the sand upward into ridges. Sand-filled cracks in dry lakes of the past have, in places, been turned into stone (been lithified), preserving for us an idea of an environment of yesterday.

The Racetrack, a playa in Death Valley National Monument, in company with one or two other playas in western United States, is the scene of a curious mystery. There, following storms, one may see the tracks of "walking" stones

FIG. 58. "Walking" stones at Racetrack Playa, near Death Valley.

on the smooth clay surface of the playa. There are no tracks of anything (or of anyone) that might have pushed or pulled them.

In a recent study of stones on The Racetrack, researchers found that, of 30 stones studied, 28 moved at least once during storms in the winters of 1968–69 and 1972–73 and 1973–74. Rocks as large as 25 kilograms (55 pounds) walked as far as 220 meters (725 feet), total, in all three winters. The longest walk during any one storm was that of a small, 250-gram (9-ounce) stone, which left a trail 202 meters (662 feet) long.

How the stones walk is not known exactly as yet. Robert P. Sharp and Dwight Carey, who have studied the walking stones of Racetrack playa, have concluded that the wind pushes them, and that it is able to do so because the clay beneath them becomes exceedingly slick in wet weather. Other researchers, working in Nevada, have suggested that when a thin film of ice forms on a playa, the wind finds it easy to scoot the stones along. But it was never cold enough for ice to form on Racetrack playa when the stones there moved during the years of the study, so ice cannot explain the stones' movement there.

Winds strong enough to move rocks are possible in the desert, giving rise to the idea that wind is the principal sculptor of the desert landscape. Not so; despite its comparative rarity, water is still the most important agent of erosion and deposition in deserts, as in most places on Earth. Wind is of comparatively greater importance in arid climes than it is elsewhere, but, even so, water still makes the larger impression.

Water works at a different pace in the desert. It does not work at a steady, grinding pace here, but as a quick flush followed by a long wait.

The rain that falls in the desert may come so swiftly it has little time to gather itself into water courses. Instead it may sweep over the desert as a sheet, carrying with it rock, sand, and clay fragments of many sizes. It moves easily and swiftly, as there is little vegetation to impede it. After the water has passed, the fragments remain as a veneer on the landscape.

Or, instead, the rain from a desert storm may find its way into channels and bear down upon the unwary. One can be camped in an arroyo—a dry watercourse—far from an active storm, and yet be overwhelmed by a wall of water from an unseen source. The Western term "dry gulched," meaning ambushed, comes from such an event.

A normally dry arroyo (called "wadi," "oued," "chapp," and "laagte" in other languages) is the river of the desert. Arroyos seem to be rivers of sand with only a suggestion of water in rare places. The Amargosa, principal stream of Death Valley, is such a river; so is the legendary Mojave, the major river in the desert of that name. Mark Twain once said of the Humboldt, a desert river in the heart of Nevada, that "one of the pleasantest and most invigorating exercises one can contrive is to run and jump across the Humboldt River until he is overheated, and then drink it

FIG. 59. Dry river in the desert. Most rivers in the desert begin bravely, but sink into the desert sand before reaching the sea.

dry." If one were to follow such a dry river headward, one would see places where the river had lost itself in the sand; other places where patches of water indicate impervious rocks beneath.

At the point where a dry tributary drops from the mountain the dry weather "stream" may look more like a dam than a watercourse. For here, likely, the wet weather stream has built an alluvial fan—a cone-shaped deposit that marks the spot where the steep desert mountains meet the flatter desert plains. Here, mudflows or flood waters, carrying a massive amount of debris from the higher slopes, are no longer able to transport their load and must drop the debris. Each flood and mudflow through the years reshapes the details of the fan, but maintains its overall rumpled cone appearance. Higher in the mountains, one can see smaller heaps of larger rocks—talus piles derived from rocks falling one at a time or suddenly breaking loose when a cliff crumbles.

FIG. 60. Desert arroyo.

FIG. 61. Alluvial fan at the base of desert mountains.

Water does its work throughout the desert, changing the mountains and carving the spectacular canyon country. From flat-lying bedded rocks, water gradually erodes pieces and parts, leaving mesas (flat tablelands) and "temples" between the eroded parts. As erosion continues, the mesas may be reduced to buttes, then to spires; and finally, perhaps, erased entirely.

The paucity of vegetation in arid lands permits steep canyons to be carved into hard rock; but in soft rock, where there is no plant life to impede water erosion, a flamboyant display of badlands can result. In regions where sediment has not yet become hard rock, and where there are torrential rains but little other precipitation, badlands can begin as a system of ridges and rills that enlarge from the bottom to yield a complex, often intricate design of great beauty. The size of grains making up the soft sediment determines the general pattern of the badlands.

Water works in quiet ways, as well as with the sudden violence of torrent and flood. In much of the California desert, the coolness of night brings an almost immeasurable film of moisture to the landscape. This moisture, by nourishing lichens that concentrate iron and magnesium, has

FIG. 62. Desert badlands.

FIG. 63. Desert water tank.

produced the black varnish so common in deserts. How long it takes to produce this desert varnish is not known exactly. One researcher thinks it takes 25 years, based on his observations, while Charles Hunt, who has studied Death Valley for many years, considers the varnish there to be as much as 5,000 years old. Perhaps both are right.

Water, working in concert with wind, produces other effects in the desert, as well. For instance, water will stand on flat surfaces longer than inclined ones. The longer it stays, the greater its weathering effect on the rock. While it is there, it may manage to weather out a few particles that the wind can then remove. In the next storm, water will seek this now slightly lower spot, and will linger even longer there than before. By this means, the spot will gradually be transformed into an indentation. Ranchers in the Southwest call these indentations "tanks" or "tinajas" ("pans" is another term) and rely on them for emergency water.

Tanks are most likely to form in shady places or in soft

FIG. 64. Desert rock honeycombed with tafoni.

spots in the rock. Even inclined rock in shady places remains moist longer than rock in sun. Any fragment loosened by lingering water drops by gravity or is removed by wind. Soft spots, too, are more vulnerable. Gradually, even inclined rock may become honeycombed with holes (called "tafoni").

There are, of course, larger holes in the desert that qualify as true limestone caves. Unlike caves in more humid climates that are still developing, most caves in the desert are essentially dead. Stalactites and stalagmites and other cave features no longer grow because of the lack of water. Unless there is a radical change in climate, the broken stalactite "collected" by the thoughtless tourist or careless scientist will never be replaced, even in the thousands of years it would take to form them in more humid climates. In California, Mitchell Caverns in San Bernardino County is such a dead souvenir of a wetter past.

Water and wind work in concert to produce another characteristic feature of arid lands: desert pavement. Most California deserts are floored by a mosaic of flat-lying stones covered with desert varnish. If one looks carefully, one can see that they rest in a matrix of fine clay. The fine clay is the clue to their origin: in storms, it becomes wet and swells, forming a slurry on which the flat stones tend to float, like raisins in a pudding. When the clay dries it shrinks, leaving the stones tightly bound to it, nearly touching one another. Any dried, loose clay left on the surface of the rocks is then blown away by the wind.

Wind. Wind strong enough to lift dust and clay thousands of feet; wind powerful enough to move rocks. It is wind that is the signature of the desert, if not its prime mover.

As a sculptor, wind is relatively ineffective. Water is the major artist, wind an assistant whose artistry is limited to minor parts of the masterpiece. Because only very strong winds can lift sand-sized particles higher than about 2 meters (6 feet), sandblasting effects are confined to about that distance above the ground. If one sees what appears to be sandblasting several meters above the Earth's surface, one should suspect that perhaps the ground itself has been lowered.

Polishing by sandblasting is one of the wind's main erosive tasks. In some places, it has grooved and polished rocks to give them a fluted appearance, like the robe of a medieval statue. Many stones in the desert have been polished and flattened by wind (and water), so that they have two or three facets at angles to one another, as cut diamonds do. These stones are called "ventifacts." The faces may mark different directions of the wind at different seasons, or they may mean that the stone has been moved.

Wind's major work in the desert is as a transporter. It picks up fine dust, silt, and clay, lofts it high in the air, and sometimes out of the desert. Since the drying of Owens Lake in this century, after its source of water was taken by Los Angeles, Owens Valley has had many serious dust storms. The particles of dust are very fine (many less than 0.1 mm in diameter), and because many of them are salts from the dry lake bed (such as sodium sulfate, sodium car-

FIG. 65. Ventifacts.

FIG. 66. Mushroom Rock, a wind-scoured basalt column.

bonate, and sodium chloride), they are a health hazard to those in nearby communities. One Owens Lake dust storm lifted more than 20,000 tons of material into the air as high as 1,200 meters (4,000 feet), creating plumes that covered 9,000 square kilometers (3,500 square miles) and were visible for 250 kilometers (150 miles)—large enough to be seen on satellite images. Dust storms like this cause darkness at noon and bring activity to a halt. Airplanes cannot land, cars cannot move. After the silent darkness has passed, one is surprised to discover that the velvet dust has abraded windshields and removed paint. Storms of dust are not confined to Owens Valley. Virtually all the southern deserts, as well as the great Central Valley and even areas along the coast, have seen the blackness come.

Twisters, or dust devils, another form the wind can take, are made visible by the dust the wind carries. Some are

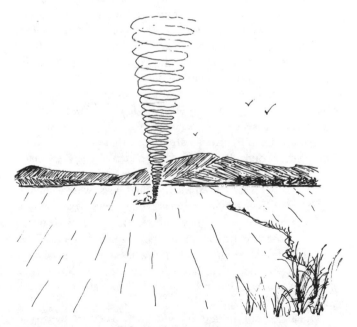

FIG. 67. A desert twister.

more than a mile high, and march across the desert like great gray funnels. They can be destructive: one huge dust devil eliminated an avocet rookery of 400 nests on Paoha Island, Mono Lake, in one quick twist; another twister destroyed a church in Tucson, Arizona, in 1964. With this kind of destructive power, it is surprising to find that some dust devils can be dispersed by merely running into them. It is wise to try this only with smaller devils, however.

Wind carries much of the dust and fine particles out of the desert–perhaps ultimately to the sea. Coarser particles, such as sand, stay in the arid land where the wind shuffles and reshuffles them, forming them into dunes of various kinds and shapes.

Dunes are an extravagance of nature. They form wherever there is a surplus of loose, dry material of the proper size and wind to flow it. Most dunes are sand, because sand is durable, but dunes can form of any material of the proper size and abundance. The proper size averages about 0.177 mm, with more than 90 percent of the grains measuring less than 0.25 mm in diameter. Coarser fragments are too difficult for the wind to move, and finer material can be whisked out of the desert in dust storms.

Clean, light-colored dunes are composed mainly of quartz sand, the grains perhaps rounded from their previous adventures in seas or on beaches, and probably frosted by abrasion (tiny silt particles are especially effective in frosting). Not all dunes are dominantly quartz. If they are made of material from a nearby source, their grains are likely to be more diverse in composition and less well rounded. Dunes in White Sands National Monument, New Mexico, for example, are made of gypsum; Barking Dunes, Hawaii, are composed of fragments of shell; dunes in Great Sand Dunes National Monument, Colorado, contain many fragments of volcanic rock derived from San Juan Mountains; dunes at Cinder Cone in Lassen Volcanic National Park are made of buff volcanic ash.

Dunes are not limited to hot deserts. They are the signa-

MAP 17. Sand dunes of California.

ture of the wind in all deserts, high and low, tropical and polar, hot and cold. Sand dunes are so characteristic of the desert (although sea shores and river beds have them, too) that one tends to visualize the desert as a vast sea of sand. It is not. The sandiest desert in the world, the Arabian, is less than one-third sand dunes. North American deserts average 1 percent; even in Death Valley, which has the world's most photographed sand dunes, they form only about 2 percent of the area.

There are, of course, vast sand seas in the world's deserts called "ergs," but none in California. The largest erg in the world is Rub al Khali, Africa, which covers more than 560,000 square kilometers (210,000 square miles). From images made by our orbiting satellites, researchers have detected that Earth's sand seas are in regular, harmonic patterns. Edwin McKee and Carol Breed have distinguished five distinct configurations of the world's ergs: a parallel, straight pattern, such as is seen in the Simpson Desert of Australia, or the Empty Quarter of Saudi Arabia; a parallel, wavy pattern, such as one can discern in the Nebraska Sand Hills; a star, or giant pinwheel pattern, as in the Great Eastern Erg of Algeria or the Sonoran Desert of Mexico; U-shaped patterns such as those at White Sands National Monument; and sheets, or stringers, as at Lima, Peru, where sand is not gathered into distinct dunes.

Smaller than ergs, but larger than individual dunes, are groups of dunes called "draa" which may have dunes on their own backs. Individual dunes, too, are of several different types. Although in English we use the word "dune" to cover all types, in Arabic and African languages there are many words for dunes, some of which we have borrowed. A long, sinuous ridge (probably formed where there is a large supply of sand and wind from one dominant direction) is called an "akle" or "uruq"; dunes that have a twisted shape like an Arabic sword are called "seif" dunes; tall, sand mountains have a star shape (see page 209), and probably form from shifting winds of equal velocity. Horseshoe-shaped (parabolic) dunes may form behind hollows, their arms surrounding the hollow and pointing windward. If a strong wind blows the center of the dune away, the arms may be left stranded as thin stringers.

One special kind of dune familiar to everyone is the sand shadow, or "coppice" dune that forms behind objects. "Nebkha" is an Arabic word for a small one; "rebdou" for a large one. The exact form of a dune depends upon the

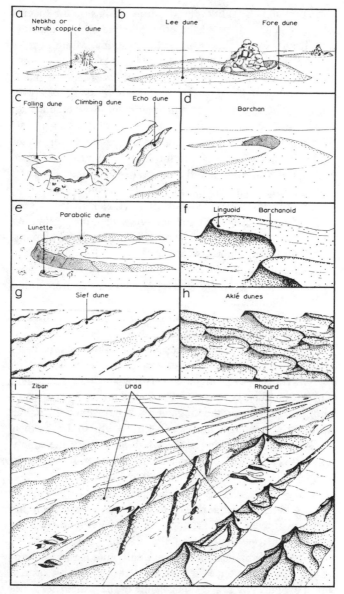

FIG. 68. Types of dunes.

direction and strength of the wind, as well as on the species of plant or the shape of the object behind which it forms. The Devils Cornfield, Death Valley National Monument, is a special case of a coppice dune. Here arrowweed plants have trapped sand to grow in. The more vigorous the plant, the more sand it traps. As it traps more sand, the plant grows higher, so that both dune and plant grow together. In some deserts, such mounds may reach 20 feet in height. The plants grow as long as their roots can reach the water table. If the mound becomes too high for the roots to reach water, the plant dies and the mound disintegrates, leaving a colored ring in the sandy soil to mark the site.

Unlike water, the wind can lift sand over obstacles, and in so doing, forms two special types of dunes: climbing dunes, which are sheets of sand being blown over a mountain; and falling dunes, which are sheets of sand that have been blown over the top. A falling dune in the shape of a cat is huddled against Cat (Cronese) Mountain, San Bernardino County, California.

Any of these dunes may form on the backs of draa; in turn, any of them may have sand ripples on their backs. Ripples are universal; they ride up and over dunes, form on their backs, and in hollows between dunes. They are not, as one might suspect, infant dunes, but are a separate sand form altogether. If a steady wind blows over a sheet of sand, ripples will form. If the sand supply is large enough and the wind strong enough, dunes will form with ripples on their backs. Dunes will grow as the wind increases if the supply of sand is adequate, but ripples will form even if the supply is sparse.

Dunes are a reflection of the unseen air. Turbulence in the air, caused by differences in pressure derived from differences in heat of the ground, variations in cloud cover, or changes in topography, alter the direction and velocity of the wind. The movement of air we call wind is three dimensional. Unlike water, it is not confined to channels, or held upon the earth by gravity. The shape of the wind determines

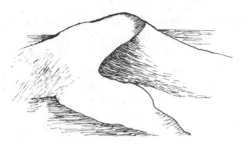

FIG. 69. Barchan dune.

the shape of dunes, and what is surprising is not that there are so many shapes, but that they are so regular.

The elegance of a sand dune is rarely surpassed in nature or in art. From a simple sand shadow that forms behind an object, through the various forms of individual dunes—stars, barchans, seif—all have a simple grace of design that delights the eye.

The dune one thinks of as most representative of the entire dune family is the barchan (pronounced bar-cane) dune, which is a neatly curved dune whose arms point away from the wind. They form where there is a moderate amount of sand and one dominant wind direction. The windward face of a barchan accepts new sand from the wind. If the wind is strong enough, it can push particles to the crest, where they may rest for a time. The angle of the face of a dry barchan dune is always about 32°. The reason for this is that piles of particles will rest at various angles, depending on the size and shape of the particles. As dunes are made of sand within a certain size range, the "angle of repose" for that size of particle is 32°. Coarser particles have lower angles; finer particles higher angles. (A fine barchan dune field stands west of the Salton Sea near State Highway 86.)

The wind moves a curtain of sand that hugs the surface of the ground, and moves still other grains by bouncing them along. Grains rarely bounce more than 50 centimeters (20 inches) high, but they knock other grains as they

bounce, causing the stuck grain to bounce, also, and frosting both of them. At times the low, moving curtain and the
bouncing grains become so thick one cannot see the ground,
yet above the 50-centimeter level, the air may be clear.

The creeping and bouncing grains are those that create
dunes. In general, the stronger the wind, the higher the
dunes build. Gusts are particularly important because the
rate of movement varies with the cube of the wind velocity.
This means that a wind of 36 miles per hour can move 3½
times as much sand in 1 minute as a wind of 24 miles per
hour, and a wind of 54 miles per hour can move 11 times as
much as a 24 miles-per-hour wind.

A "smoking" dune shows quite plainly that it is in motion. Sand, streaming up the windward face, forms an aerial
fringe as it plunges up and over the top to hurtle down the
slip face. If too much sand collects on the crest, thus exceeding the angle of repose, patches of sand break loose to
avalanche down the lee slope.

Edwin D. McKee, long a student of sand dunes, watched
dunes change several seasons at White Sands National
Monument, New Mexico, and clocked their movement at
an average of 38 feet per year. Although this does not seem
to be very fast, as planes fly or men walk, it is inexorable.
Where roads and towns have been built near these dunes,
the dunes have covered some of the roads and buildings,
rendering the roads impassable and the buildings useless. In
the western part of Afghanistan, along the Iranian border,
where shifting sand covers one's tracks as rapidly as they
are made, geologic field parties always go in convoy—two
or more off-road vehicles together, with never fewer than
five men. "Too dangerous," said Afghan geologist S. M.
Gazanfari. "A jeep can be buried while we're away for a
few hours of fieldwork. A man alone with no transportation
or supplies in sand seas like these cannot last long."

Despite this constant movement, dunes may persist for
thousands of years in the same spot, or near the same spot,

perhaps moving one direction in winter, only to return to the original spot when summer comes.

If one is around active dunes long, they begin to seem alive, and to have a power and personality of their own. They move secretly and silently, reforming, yet maintaining their general shape for millennia. Nothing but a heap of sand? They often do not seem so.

Dunes adjust quickly to changes, magnifying footprints of large animals, or neatly retaining the tracks of nearly weightless insects. Yet these can be quickly erased as a dune makes its adjustments, leaving a pristine dune that appears as untrodden as on its day of creation.

When humans feel that, for their own purposes, the progress of a dune must be halted, heroic measures are often required. Dunes can be "killed" by cutting off the wind and sand, or by keeping the grains from moving. How can one cut off the wind and sand? One way is to erect barriers. The dust storms of the Midwest are now lessened by a shelter belt of trees. Or, fences can protect roads—at least until the fence is overtopped by a dune built up at its foot.

Dunes are prevented from moving naturally by the growth of plants on them, by a change in the pattern of the wind, or by a change in the climate. Some desert dunes, if an adequate supply of water becomes available either naturally or artificially, can be anchored by plants. Sometimes, however, as in the case of the Death Valley dunes, the plants one sees are not anchoring the dunes, but are being overrun by them. Dunes can also be artificially anchored (and uglified) by pouring oil or concrete on them.

Dunes that sing, bark, or boom have been known for more than a thousand years. Marco Polo heard some in China; travelers of long ago in the deserts of Afghanistan told of one dune that sang in Alexander's time. "Musical" dunes are known throughout the world; the Kelso dunes in the Mojave Desert are such. The noise they make has been compared with drums, bells, telephone wires, and gunfire. All dunes do not "speak" or "sing"; those that do, speak rarely—when their surface is disturbed and weather conditions are just right. Strong winds can cause dunes to boom of their own accord, or the pressure of someone walking on them may start the music. Islamic legend says that musical dunes in Arabian lands sing only on Friday (Islamic holy day), and that sand removed from them will return to the dune by nightfall. Legends from the Sinai Peninsula tell of buried monasteries tolling ghostly bells and beating spirit drums.

But what exactly causes this voice of the desert is not wholly known. The shape, size, and arrangement of the sand grains have some influence. Whatever the cause of the music, it is also played on the moon.

Singing or mute, the California desert is in danger of being destroyed. There are a few protected areas, such as Death Valley National Monument, Joshua Tree National Monument, and Anza-Borrego State Park, but the remainder of the desert is rapidly being devastated. Subdivisions, complete with plastic lakes, or "ticky tacky little houses" set like checkers on a board, have eaten up much of what could have been cherished open space.

Highways, power lines, canals, and reservoirs take up much more space, while agriculture (probably the least offensive) uses abundant acreages. Plant thieves steal the natural cover of the desert, while cruel and thoughtless gun handlers and drivers kill its animals. Even the government, which is sworn to protect it, does not do so. The policy of destroying 500-year-old juniper trees, of tearing up the sweet-smelling sage on public land to plant grass (which probably will not grow) is to be deplored. Such changes in the natural vegetation will probably lead to accelerated erosion, to the buildup of alkali layers in the soil, and to the ultimate destruction of what thin desert soil there is.

Off-road vehicles (motorcycles, 4-wheel-drive wagons, dune buggies) jeopardize the desert's very existence. By tearing up the natural soil, by deliberately and heartlessly running over its plants and animals, they expose the desert to destruction by the fierce climatic forces that already operate there. For example, in the Barstow to Las Vegas motorcycle race held in 1972, cycles lifted 600 tons of dust into the air high enough to cause a plume visible from our orbiting satellites. This is more than ten times the dust raised during a severe Owens Lake dust storm. Besides allowing the dust (a needed constituent of the soil) to be blown out of the desert, the cycles have destroyed desert pavement that took centuries to form and that protects the desert from further damage. Plants and animals in a wide swath around the race course were destroyed. The desert will be more than a century recovering—if it recovers at all. Surely, destruction of this caliber can scarcely be termed "legitimate" use of the desert!

Yet off-road vehicles are a great boon to many of us who would like to seek the quiet of the open spaces that are left. Wisely used and thoroughly controlled, the off-road vehicle is a blessing. Allowed to roam where its driver wills, it is a clear and present danger.

Once, the western desert was a land where travelers suffered and sometimes died; today, it is the desert that suffers and well may die.

TABLE 14. Features of the Desert

Type of feature	Feature	Where to see a good example
Independent of climate	Fault	(see table 12 and map 10)
	Hot spring	see table 8
	River, throughgoing	Colorado River, southern California
	Volcanic feature	see table 5
Inherited	Cave	Mitchell Caverns State Park
	Dry watercourse	Mojave River, in places
		Amargosa River, in places
	Dry waterfall	Fossil Falls, Inyo County (near Little Lake)
	Former lake level	Mono Lake
		Owens Lake
		Death Valley (Lake Manly)
		Lake Searles, Slate Range
		Lake Cahuilla, Salton Sea
		Travertine Point, Anza-Borrego Desert State Park
		Lake Manix, about 50 kilometers (30 miles) east of Barstow
	Relict lake	Mono Lake
		Eagle Lake
		Honey Lake
	Tor	Alabama Hills
		Joshua Tree National Monument
		Mojave Desert
		Box Spring Mountains, San Diego County

TABLE 14. Features of the Desert (Contd.)

Type of feature	Feature	Where to see a good example
		State Highway 15 near Escondido
	Tufa domes	Mono Lake
		Honey Lake
		Searles Lake (Trona Pinnacles)
Water deposited	Alluvial fan	Black Mountains, Death Valley National Monument
		Panamint Mountains
		Copper Canyon, Death Valley National Monument
		Panamint Butte, Panamint Valley
		Wineglass Canyon, Death Valley National Monument
		Hanaupah Canyon, Death Valley National Monument
		Milner fan, White Mountains
		Coachella Valley
		San Jacinto Mountains
		Mojave Desert
		Bullion Mountains
		El Paso Mountains, near Railroad Canyon
		Rand Mountains, near Randsburg
		Owens Valley
		White Mountains east of Bishop
		Inyo Mountains, near Lone Pine
		Palm Avenue, San Bernardino city

TABLE 14. Features of the Desert (Contd.)

Type of feature	Feature	Where to see a good example
		Palm Springs area
		West side of San Joaquin Valley
	Caliche	Many places in desert
		(also in Corral Hollow, Alameda County)
	Gravel bar (in an ancient lake)	Death Valley
		Near Hot Mineral Spa (Lake Cahuilla)
		Lake Manix east of Barstow
	Mudflow	Death Valley National Monument
		Many places in desert
		See also table 9
	Patterned ground	
	Giant	Airport playa
	Normal	Most clay playas see map 18
	Salt polygons	Death Valley
		Salton Sea near Obsidian Butte
	Playa	See map 18
	Playa:	
	Clay surface	The Racetrack, Death Valley National Monument
	Salt surface	Death Valley
	Self-rising ground	Many playas—see map 18
	Tombolo	Ancient Lake Cahuilla at Travertine Point, Salton Sea

TABLE 14. Features of the Desert (Contd.)

Type of feature	Feature	Where to see a good example
	Talus cone	Death Valley National Monument
Water eroded	Badlands	Death Valley National Monument
		Anza-Borrego Desert State Park
		Red Rock Canyon State Park
		Carrizo Plain
		Fish Creek Mountains
		Ibex Pass, near Shoshone
		Coachella Valley
		Railroad Canyon, State Highway 14
		Tecopa, near Death Valley National Monument
	Canyons	Titus Canyon, Death Valley National Monument
	Pediment	Silver Mountain, north of Victorville
		Black Mountains, Death Valley National Monument
		Lucerne Valley
		Red Rock Canyon, State Highway 14
		South face of Granite Mountains between Kelso and Amboy
		Northeast side of Old Woman Mountains
	Tank	Joshua Tree National Monument
Wind deposited	Sand dunes	See map 17

TABLE 14. Features of the Desert (Contd.)

Type of feature	Feature	Where to see a good example
Wind eroded	Grooves, fluting	Aeolian Buttes, near Mono Lake
		Garnet Hill, Coachella Valley
		Hidden Valley, Joshua Tree National Monument
	Tank (pan)	Joshua Tree National Monument
	Ventifact	Many desert pavements, Death Valley National Monument
		Garnet Hill, Coachella Valley
		Aeolian Buttes, Mono County
		Amboy lava field, San Bernardino County
		San Felipe badlands
	Yardang	Danby Dry Lake
		Rogers Dry Lake
		San Felipe badlands
Formed by wind and water	Desert pavement	Death Valley National Monument
		Searles Lake
		Anza-Borrego State Park
		Joshua Tree National Monument
		Salton Sea area
		Amboy lava flow, San Bernardino County
		Victorville, 12 kilometers (7 miles) north on Interstate Highway 15
		Cargo Muchacho Mountains, Imperial County

TABLE 14. Features of the Desert (Contd.)

Type of feature	Feature	Where to see a good example
	Desert varnish	Death Valley National Monument
		Alabama Hills
		Joshua Tree National Monument
		Anza-Borrego State Park
		Black Mountains, Red Rock Canyon, State Highway 14
		Coachella Valley
	Fantastic weathering	See table 10
	Sliding stones	The Racetrack, Death Valley National Monument
	Tafoni	See table 8
Cold deserts	Patterned ground	White Mountains

10 · THE HUMAN LANDSCAPE

More than two centuries ago, a small band of Franciscan missionaries started their way along the unknown coast of Alta California. They chose sites for missions as they went, selecting places that had good water and agricultural possibilities. They chose cleverly and well; the missions they and their successors established became the nuclei of urban California.

No doubt the venerable fathers would have been startled if they could have foreseen the results of their labors. From these centers, the modern phenomenon called "urbanization" has transformed what were once peaceful valleys into a new landform feature: the city or "urb." Mountains have been removed to make new stone with which to erect buildings, build roads, and cover the land; forests have been erased by axe and chain saw; wild creatures have been killed or crowded from their homes. Lakes dot the landscape where none were before; in some places, lawns and trees grow on little flat spots in desert mountains; in other places, rows of green crops supplant stony desert pavement.

In the years following the founding of the missions—indeed, until very recently—we looked upon the sweeping changes wrought by humans on Earth as either good or negligible. In good conscience, we bragged about draining the swamps and the civilizing of wild places. We earnestly believed our changes were "progress," little realizing that many of our engineering marvels are clumsy at best and lethal crimes at worst.

Some human changes are—at least in human terms—good for humans in the short run; they often are not good for the rest of life on Earth, and may not be good for humans in the long run. In terms of landforms, the past two centuries have seen the most rapid changes California has ever known.

Someone has likened the city to a large organism that eats, breathes, and discharges wastes. Looked at in this way, it is a large, living landform that grows at the expense of its surroundings. Food and water are brought in for the organism: food that we may have grown by deforesting our hills and denuding our prairies; water that we have captured and moved. In addition to food and water, a constant stream of hard materials for the construction of the "bones" of the city pour into it. These include rock, sand, and gravel from quarries, together with cement manufactured from other quarried rocks; iron, steel, and metals from mines near or far; asbestos in a variety of guises from mines and quarries all over North America, as well as a host of other mineral materials, a large portion of them derived from the chemical rearrangement of petroleum. In the winning of all these, the landscape has been radically altered.

By these, and by the many others that pour into the city, the city grows. As it grows, it spreads its tentacles in all directions—as roads and highways, rails and tunnels, pipes and drains. Each road, each pipe, each tunnel alters the landscape, and in altering, creates a new set of possibilities for further change.

It is not true that the quantity of waste products that is moved or spilled out of the city is larger than the quantity going in, but because of the disarray in which the waste is removed, it often seems so. The waste products are peculiarly hard to manage. We can no longer simply throw them outdoors, as in primitive days and places; neither can we

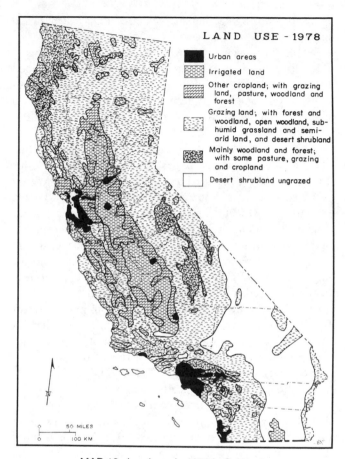

MAP 18. Land use in 1978 in California.

dump them in our water supply as has been our habit; burning is not always a good solution as it befouls the air, and is wasteful of reusable materials. Many cities have used the holes dug for minerals in which to bury the mounting trash, but this is not always good, either, as it may defile the groundwater. (In addition, we have the special problem of deadly nuclear waste and stocks of incredible military and industrial poisons that have been built up and that are very difficult to dump safely.)

So as we cut down mountains so the city can grow, we make new ones of its ordure.

The growth of cities and of the population in general has had a profound effect on the landscape, and may have an increasingly great effect as time goes by. In our need for additional water supplies, we have left but a handful of streams undammed. The dams themselves are significant landforms, and the reservoirs they create are major landscape alterations. Large dams may cause further landscape changes by triggering earthquakes. All dams collect sediment, and will, in time, be flat valleys unless somehow they are heightened or their reservoirs excavated.

The sediment they catch is what once would have gone to build sand bars and beaches. In its absence, sand beaches are slowly withering to a collection of stones. In some places, where sea walls and groins have been built to protect beaches, the walls themselves change the pattern of sand movement so that the beaches are lost.

The water we are saving behind dams is used for industry, agriculture, and in swelling cities. In southern California, the equivalent of 130 centimeters (50 inches) of rain—as much as falls on the rain forests of northern California—is poured over lawns each year. The lawns and the buildings with them are carved from desert hillslopes that are kept from sliding into the valleys only with difficulty. Indeed, there is presently no way of preventing many slides; we must, instead, learn to live with them. In our eagerness for water, we have seeded the clouds to change the climate, and in so doing, have washed away part of the landscape beneath.

In our haste to build on steep slopes (where the view is better), or to obtain lumber for building, we have ripped up our forests, increasing the possibility of slides and accelerating the rate of erosion. In Redwood National Forest, where lumbermen have cut redwoods thousands of years

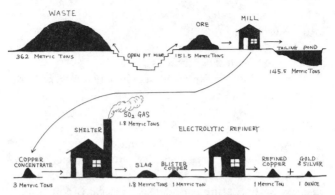

FIG. 70. Flow sheet showing how much waste is generated from the production of a ton of copper. The average copper mine today "contains about 0.6 percent recoverable copper. About 362 metric tons of ore will have to be removed for each ton of copper extracted, thereby leaving a hole in the ground as big as a small house (7,400 cubic feet) and an equivalent pile of waste rock on the ground nearby.

"The 151.5 metric tons of ore-bearing rock taken from the mine are brought to a mill where they are finely ground to separate the metallic minerals from the worthless matrix, or gangue. This process will produce almost 3 metric tons of concentrate and 145.5 metric tons (2,800 cubic feet) of finely ground rock, or tailings, which must be deposited in settling ponds to prevent silting in natural waterways. The mill operation requires 90,000 gallons (341,000 liters) of water, up to two-thirds of which is recycled, and 3,340 kilowatt-hours of electricity, equivalent to almost half a metric ton of coal.

"The concentrate, consisting primarily of copper and iron sulfide minerals, now goes to the smelter where it is mixed with limestone and quartz sand. The smelting process is complicated, but essentially it burns off the sulfur in the concentrate, removes the iron as a glassy slag, and produces molten, impure blister copper. With each metric ton of copper thus resulting, approximately 1.8 metric tons of slag are produced, and 0.9 metric ton of sulfur is burned, which yields 1.8 metric tons of sulfur dioxide gas. Part of the sulfur dioxide may be converted to sulfuric acid; usually 2.7 metric tons of acid are produced for every 1.8 metric tons of sulfur dioxide gas, but some of the gas is vented into the atmosphere. In smelting this metric ton of copper concentrate, about 22 million BTU's (5,544,000,000 calories) of heat are consumed, an amount equivalent to 22,000 cubic feet of natural gas or almost 1 metric ton of soft coal.

"Further treatment of the 1 metric ton of blister copper in an electrolytic refinery yields pure copper and about $35 worth of gold and silver as a by-product. This last process requires roughly 1,250 kilowatt-hours of electricity."

old, this rate of erosion has increased by perhaps as much as 70 times, making it the highest in the nation. This rate is 32 percent greater than the rate of the Eel River, which is wearing down the Klamath Mountains at the nation's highest rate.

All of this literally throws away the soil. California originally had about 8,700,000 acres of land with soil of the highest quality. Today there are only 7,000,000 acres left unburied by buildings or roads, and 7 percent of that—500,000 acres—is zoned so that it could be covered by developments in the next decade. In addition, as a result of logging, construction, farming, and flood-control practices, 60,000 acre-feet of soil are lost through erosion each year.

We are wearing out the land where we are not covering it up. We have learned a little about wise agricultural practices, but we have yet to learn a valuable lesson regarding our uncontrolled play. In the name of recreation we are destroying much of our irreplaceable inheritance. Off-road vehicles, for example, have damaged agricultural soil as well as the exceedingly fragile soil of the desert so greatly that much of it cannot be repaired within two centuries; in other places, the soil has been worn away completely, so that millennia may not suffice to replace it.

We are so accustomed to the changes we ourselves have made that it is hard for us to imagine California in its original, prehuman state. We can discover how it was two centuries ago, thanks to historic documents. Here is a description of the San Francisco Bay area as reconstructed by D. W. Mayfield, who wrote in *Pacific Discovery*:

> Two hundred years ago the eastward view from San Francisco's Twin Peaks—looking across the bay—would have shown most of the Oakland and Berkeley Hills practically devoid of woody vegetation. The flats in front of these hills were also dominated by grasses and herbs, except where strips of riparian vegetation shrouded meandering streams and where tangles of tule and willow—signs of freshwater marshes—bordered the pickleweed-dominated salt marshes. [There was] a single dark green grove of

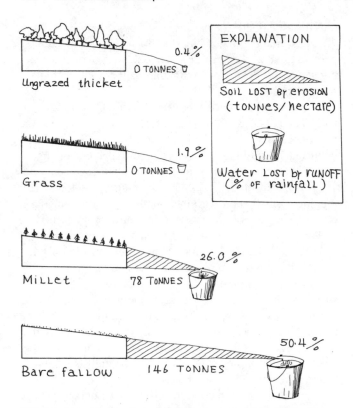

FIG. 71. Results of soil erosion tests on ground with different types of plant cover in a semiarid part of Tanzania. The ungrazed thicket and grass lost little soil, whereas bare ground lost a great deal of soil.

oaks on the Alameda Peninsula, and the clustered, sharp silhouettes of redwood trees crowned an inland peak behind Oakland.

Low hills and valleys, now under the buildings of downtown San Francisco, were primarily covered with grasses and coastal scrub. The North Beach district was, as the name implies, a sandy beach, not like the muddy marshes seen surrounding most of the bay. A lagoon surrounded by willows, cattails, tules, and pickleweed—all teeming with waterfowl—once submerged lands now under the Mission district. Potrero Hill, to the east, was practically an island in the marshes.

North and west of Twin Peaks, grassy patches and brushy growth colonized hundreds of acres of dunes, with some small live oaks clustered in wind-sheltered places. (Here, expanding San Francisco destroyed one of California's largest dune complexes.) San Bruno Mountain, a few miles southeast from Twin Peaks, could have been viewed two centuries ago in much the same grassy condition that it is today.

In the vast alluvial floor of the Santa Clara Valley, ground water was so plentiful before historic exploitation that many areas were impassable during the wet season, when the water table was literally at the surface. Many permanently wet areas created huge freshwater marshes that were variously dominated by tules, cattails, willows, and sometimes alder. The largest of these marshes were located along the Guadalupe River, in and near Santa Clara, to the northwest at Moffett Field, and also in the East Bay area, at the western end of Livermore Valley. By the end of the nineteenth century, groundwater drawdown for irrigation and urban uses almost completely obliterated the freshwater marshes.

Over two centuries have passed since the first Spanish land expedition encountered the San Francisco Bay area. Land once covered with grasslands, marshes, and oak savannas is now under many different uses, and its former wildlife is mostly eliminated. But the landscape as it was [two hundred years ago] was not a pristine wilderness. It was a discovered land only to the colonizers and their civilization. To its indigenous peoples it was a home for several thousand years, a land . . . that [served] their way of life.

This land of marshes, savannas, and grasses has given way to concrete and glass, to telephone poles, roads, and skyscrapers. The wildlife has been killed, the groundwater drained—even the soil has been concealed.

Many books have been and will be written about how we humans change the Earth. This brief summary does not begin to show the shape of the changes wrought by this new natural agent, humanity. For we are indeed agents of erosion and deposition. We change the landscape radically, and cause other changes to take place. We move plants from one locality to another; we propagate new ones; we change the composition of the local fauna by introducing new, strange species and killing native ones, or by crowding them out of their homes.

By poisoning the land and water, we speed natural selection (or rather, "unnatural" selection) encouraging species that can cope with our poisons, and eliminating those that cannot. Our activities change even the air itself, causing smog, adding chemicals and particulates, raising the temperature, perhaps blanketing ourselves in a perpetual haze. Some of these changes make the rain into a chemical that can even more quickly erode the landscape.

It is possible that, by changing our air in the many ways we do, we are helping to bring about a new ice age. That, in turn, could change the relationship of land and sea, drying areas now underwater, and perhaps later flooding low-lying lands. If this occurs, we will have contrived landscape changes beyond our present imagining.

Glossary

Aa   A blocky form of lava.

Active fault   A fault that has moved in historic time, or that has had recurrent seismic movement (see *Fault*).

Akle   A sand dune in the shape of a long, sinuous ridge (also called uruq).

Alluvial fan   A low fan-shaped, sloping mass of loose rock material deposited by a stream where it issues from a narrow mountain valley, or where the stream abruptly changes gradient.

Andesite   A dark-colored volcanic rock, consisting mainly of feldspar.

Anticline   A fold that is convex upward (see also *Fold*).

Arid   Said of a climate characterized by dryness.

Arroyo   A small, deep gully or the channel of a stream that runs occasionally, but is usually dry (also called wadi).

Asphalt pit   A concentration of natural asphalt formed in oil-bearing rocks by the evaporation of volatile constituents.

Avalanche   A mass of snow, rock, or rubble that slides or falls very rapidly (see also *Landslide*).

Badlands   Rough, deeply gullied topography.

Barchan dune   A crescent-shaped dune with arms (horns) that point downwind. The windward face is gently sloping, the lee face inside the horns is steep.

Barrier beach   A long, narrow beach built by offshore wave action, separated from the mainland by a lagoon or bay.

Basalt   A dark-colored volcanic rock.

Bergschrund   A deep crevasse at the head of a mountain glacier.

Bomb, volcanic   See *Volcanic bomb*.

Braided stream   A stream made of many channels that branch and rejoin.

Caliche   A hard soil zone, made of calcium carbonate, that forms in desert climates.

Carbon-14 dating   A method of determining the age in years of an object by measuring the amount of radioactive carbon remaining in once-living material.

Cinder cone   A conical hill formed by the eruption of sand-sized particles of lava.

Cirque   A bowl-shaped depression at the head of a glacial valley.

Col   A high, narrow pass in a mountain range between two mountain peaks formed by the intersection of two cirques.

Composite volcano   See *Stratovolcano*.

Concretion   A hard, rounded mass formed in sedimentary rock.

Cone, cinder   See *Cinder cone*.

Continental shelf   The flat offshore strip of ocean bottom between the shoreline and the continental slope marking the edge of a continent.

Continental slope   A steeply dipping slope between the ocean basin and the continental shelf.

Coppice dune   A mound of sand gathered around a shrub.

Crag-and-tail   See *Roche moutonnée*.

Crater, volcanic   See *Volcanic crater*.

Crevasse   A crack in a glacier.

Creep   The slow movement of the Earth's crust along a fault, or of snow, rock, or soil in response to gravity.

Cyclopean stairs   See *Glacial stairway*.

Debris flow   A thick mud flow containing coarse-grained materials.

Delta   A low, nearly flat alluvial tract of land formed where the mouth of a stream empties into a body of water.

Dendritic drainage pattern   A treelike pattern made by streams and their tributaries branching from a master stream.

Dome, volcanic   See *Volcanic dome*.

Draa   A group of dunes.

Drowned valley   A stream valley that has been inundated by the sea.

Dune   A low bank of granular material (usually sand) with a characteristic shape, capable of being moved by the wind from place to place.

Dust devil   A relatively small, twisting windstorm.

Ephemeral lake   A short-lived lake.

Erg   A sand sea.

Erratic   A rock fragment carried by a glacier and deposited at some distance from its parent rock.

Extrusive   Igneous rock that has cooled on the surface of the Earth (volcanic rock).

Fault   A break in the Earth along which movement has taken place (see also *Active fault*, *Fault scarp*).

Fault scarp   A cliff formed along the exposed surface of a fault.

Fault line scarp   A cliff formed by the more rapid erosion of softer rock on one side of a fault.

Feldspar A group of minerals containing aluminum and silicon together with other elements; the most widespread of all mineral groups. Feldspar is the source of clay.

Firn A substance transitional between snow and ice.

Fjord A glacially carved valley floor drowned by the sea after the glacial ice melted.

Floodplain A strip of level land alongside a river flooded in times of high water.

Flour, glacial See *Glacial flour*.

Fold A flexure in rocks resulting from earth movements (see also *Anticline* and *Syncline*).

Fumarole A volcanic vent from which gases emanate.

Giant's staircase See *Glacial stairway*.

Glacial flour See *Rock flour*.

Glacial outburst flood See *Jökulhlaup*.

Glacial polish A smooth and shiny rock surface produced by glacial action.

Glacial stairway A glaciated valley whose floor has been carved into a series of steps.

Glacial step lake A lake occupying one of the steps in a glacial stairway (see also *Glacial stairway*).

Glacier A large mass of ice that survives from year to year and moves under its own weight.

Global tectonics Earth movements involving large segments or plates of the Earth's crust.

Gouge mark A crescentic mark produced by the action of glacier ice.

Graben An elongate, downdropped area bounded by faults on its long sides; usually becomes a valley.

Granite A light-colored, plutonic igneous rock consisting of quartz and feldspar, sometimes with dark minerals. Granitic rocks have a pepper-and-salt appearance, because the mineral grains are large enough to see with the naked eye. Quartz, the glassy mineral, and feldspar, which resembles porcelain, are the "salt"; the dark grains (most commonly the dark mica called *biotite*) are the "pepper."

Great Ice Age The Pleistocene epoch of the geologic time scale (see also *Pleistocene epoch*).

Hanging valley A tributary glacial valley whose mouth is at a high level compared to the main trunk glacier, owing to deeper erosion by the main trunk glacier.

Hardpan A hard layer of soil just below the surface, produced by cementation of soil particles by relatively insoluble materials. This is the "alkali layer" in deserts; also called "duricrust" and, in the Southwest, *caliche* (see also *Caliche*).

Horn   A sharp-pointed, steep-sided pyramidal mountain peak formed at the intersection of the cirques of three or more mountain glaciers.

Hornito   A spatter cone on a lava flow.

Horst   A block of the Earth's crust that has been uplifted relative to the land on either side. A horst is elongate and bounded by faults (compare *Graben*).

Hydrologic cycle   The water cycle, or the constant circulation of water in various forms through the atmosphere, the land, and the sea.

Igneous   Said of a rock that has solidified from a molten or nearly molten state. One of the three great classes of rocks (compare *Sedimentary* and *Metamorphic*).

Intensity   A measurement of the relative amount of damage done by an earthquake (compare *Magnitude*).

Intrusive rock   Rock that has, in a molten state, forced its way into preexisting rock (see also *Plutonic*).

Isotope   One of two or more species of the same chemical element having slightly different atomic weights. Isotopes of an element have the same chemical properties, but differ from one another physically. Radioisotopes (that is, isotopes that are radioactive) are useful in calculating the age of certain objects (including rocks) in years.

Joint   A surface along which a rock may break or part. Joints commonly form in three directions, at about right angles to one another.

Jökulhlaup   A glacier outburst flood.

Lagoon   A shallow stretch of water separated from a larger body of water or from the open ocean by a reef, barrier island, or a spit.

Lahar   A volcanic mudflow.

Landslide   A general term embracing the slow or fast movement of rock or soil down a slope by gravity (see also *Avalanche* and *Debris flow*).

Lapilli   Fragments blown from a volcano that are sand to gravel in size.

Latite   A volcanic rock close to andesite in composition and color (see also *Andesite*).

Lava tubes   Hollow spaces or caves that form under the cooled crust of a lava flow.

Lichenometry　The technique of using lichens growing on rocks to determine the length of time of certain events in landscape changes.

Lithify　To harden to stone.

Little Ice Age　See *Neoglacial*.

Maar　A volcanic crater often filled with water.

Magma　Molten rock material, from which both volcanic and plutonic igneous rocks have solidified. Molten rock that is deep underground is termed magma; when it reaches the surface (as in volcanic eruptions), the fluid rock is called lava (see also *Lava*).

Magnitude　A measure of the strength of an earthquake (compare *Intensity*).

Mantle　The zone of the Earth's interior below the crust and above the core.

Matterhorn　See *Horn*.

Meander　One of a series of regular sinuous curves or windings in a stream course.

Mesa　An isolated, nearly level landmass standing above its surroundings and bounded by steep cliffs. A mesa has a cap that is more erosion resistant than the rock beneath. Also called table mountain. (The word *mesa* is Spanish for "table.")

Metamorphic　An adjective applying to the process of metamorphism; one of the three great classes of rocks (compare *Igneous* and *Sedimentary*; see also *Metamorphism*).

Metamorphism　The adjustment of solid rocks to changing physical and chemical environments. Commonly, rocks adjust by realigning their elements to form new minerals. Metamorphic rocks are changed rocks; they change while in the solid state, usually in response to elevated heat or pressure or both (compare *Igneous* and *Sedimentary*).

Montmorillonite　A swelling clay mineral.

Moraine　A mound or ridge of till deposited by a glacier (see also *Till*).

Mountain glacier　A glacier formed at a high elevation on a mountain slope or in a mountain valley (also called valley glacier or alpine glacier).

Natural arch　A natural landform shaped like an arch; called such in the Southwest if it has no river or stream flowing through it. If a stream is flowing beneath it (or once did), the feature is called a *natural bridge*. In California, many of the arches and bridges were formed along the coast by wave erosion.

Natural bridge See *Natural arch*.

Neoglacial The advance of ice during the Little Ice Age following the Great Ice Age.

Nueé ardente A "glowing cloud" of hot gas and fragments produced in some volcanic eruptions, which is fast moving and often incandescent.

Obsidian Dark-colored volcanic glass.

Offshore bar A sand bar, usually submerged, that serves as protection to a lagoon or harbor (compare *Barrier beach*).

Pahoehoe A ropy form of lava.

Pan A shallow, natural depression that sometimes holds water (also called a tank).

Paternoster lakes One of a chain of lakes at successive levels in a glacial valley, connected by streams or waterfalls. From a distance, the chain is reminiscent of beads on a rosary, hence its name (see also *Glacial stairway*).

Playa A dried desert lake bed (also called a *sebkha*).

Pleistocene The epoch of geologic time immediately preceding the present (Holocene) one, during which the Great Ice Age occurred. One of the two epochs of the Quaternary Period. The Pleistocene Epoch began 1 to 3 million years ago and lasted until about 10,000 years ago.

Plunging breaker A breaker that rises to a peak and breaks suddenly. Plunging breakers form where the underwater slope is steep and smooth.

Plutonic Referring to igneous rocks formed at great depth in the Earth. Many plutonic rock bodies intrude older, preexisting rocks and are therefore said to be intrusive. Granite is a plutonic rock (see also *Intrusive*).

Pluvial An adjective used to describe a feature, process, or event resulting from rain.

Pocket beach A small, narrow, crescentic beach between rocky headlands open toward the sea.

Pothole A rounded, pot-shaped pit or hole.

Pumice Volcanic rock containing many air spaces. Pumice is the hardened froth from volcanic eruptions. It is so light in weight that it will float on water.

Pyroclastic Referring to volcanic rock exploded from a volcano as distinguished from lava that issues quietly, or flows. *Volcanic ash* consists of very fine pyroclastic fragments, or clasts. Pyroclastic fragments may range in size from ash to volcanic bombs the size of houses (see also *Tephra*).

Radioactive-carbon dating See *Carbon-14 dating*.

Rhyolite A light-colored, silica-rich, iron-poor volcanic rock.

Roche moutonnée A lump of bedrock that has been sculptured by a glacier so as to have a steep, rough lee (downstream) side and a smooth stoss (upstream) side. Yosemite National Park has many roches moutonnées.

Rock flour Finely ground rock pulverized by a glacier (also called glacial flour and glacial meal).

Sea arch See *Natural arch*.

Sedimentary Referring to the process of accumulating, or transporting, weathered rock fragments (sediment) by water, air, or ice. Also refers to the rocks formed by this process. One of the three great classes of rocks (compare *Igneous* and *Metamorphic*).

Seif dune A long, sharp-crested, tapering dune. In plan view it resembles a twisting Arabic sword, hence its name.

Seismic sea wave See *Tsunami*.

Shield volcano A flattish volcano formed by successive flows of fluid lava. The shape resembles a shield.

Shingle Coarse, flat boulders and cobbles on a beach, generally with little or no sand. On some beaches, the boulders are stacked by the waves so that they overlap slightly, as shingles do on a roof, hence the name.

Slickenside A smoothly polished, striated surface on a rock resulting from movement along a fault.

Solfatara A type of fumarole, commonly sulfurous (see also *Fumarole*).

Spilling breaker A breaker whose crest collapses gradually for a long distance, the water spilling down the wave front. Spilling breakers are the best for surfing (see also *Plunging breaker*, *Surging breaker*).

Spit A fingerlike extension of beach or land projecting from the shore.

Stalactite A conical pendant hanging from the roof of a cave, resembling an icicle. Stalactites in limestone caverns are usually made of calcium carbonate; those in lava tubes are made of lava (compare *Stalagmite*).

Stalagmite A conical pendant rising from the floor of a cave. Stalagmites in limestone caverns are usually made of calcium carbonate (compare *Stalactite*).

Steppe A grassland area, somewhat drier than a prairie.

Stratovolcano A steep-sided volcano constructed of alternating layers of lava and pyroclastic material (also called composite volcano).

Stream piracy The diversion of the headwaters of one stream by a stronger, lower stream (also called stream capture).

Subduction zone A zone along which one large crustal plate of the Earth descends beneath another (see also *Global tectonics*).

Surging breaker A breaker that rises to a peak without spilling or plunging (see also *Plunging breaker*, *Spilling breaker*).

Syncline A concave upward fold in the rocks of the Earth in which the youngest rocks are in the center (compare *Anticline*).

Tafone One of the natural cavities in honeycomb structure in rocks. Plural is tafoni.

Talus Rock fragments at the base of a steep slope.

Tank A shallow, natural depression that sometimes holds water (also called a pan).

Tar pools See *Asphalt pits*.

Tarn A small lake, many created by glacial action.

Tephra A general term for all of the pyroclastic material exploded from a volcano (see also *Volcanic ash*, *Volcanic bomb*, and *Pyroclastic*).

Terrace A gently inclined surface bounded by steep slopes. A stream terrace may form along the sides of a river and a marine terrace at the edge of the sea.

Thalweg A line tracing the lowest points in a stream.

Thermoluminescence The emission of light by a substance during or after heating, at a temperature lower than required for incandescence.

Tidal wave A misleading term for a seismically generated sea wave (see *Tsunami*).

Till Unsorted glacial drift, composed of rocks of various sizes and shapes.

Tombolo A sand or gravel bar connecting an island with the mainland.

Tombstone rock Outcropping of fingers of rock eroded from nearly vertical layers. A group of tombstone rocks has the aspect of a deserted graveyard (also called gravestone rocks or gravestone slates).

Tor A pile of jointed and rounded rocks.

Tree-ring chronology The study of the annual growth of trees in order to date the past (also called dendrochronology).

Tsunami A seismically generated sea wave, confusingly called a tidal wave.

Tufa dome A mound formed of calcium carbonate (limestone) by the action of hot springs, algae, or bacteria.

Tuff Consolidated volcanic ash.

Uruq   See *Akle*.

Valley glacier   See *Mountain glacier*.

Varve   A thin bed of sediment deposited annually in a body of a water. A lake varve usually consists of two parts: a light-colored summer layer and a dark-colored winter layer.

Ventifact   A faceted pebble worn or polished by wind-blown sand.

Vernal pool   A pond or pool containing water only in the wet season.

Volcanic ash   Very fine material exploded or blown from a volcano (see also *Tephra* and *Pyroclastic*).

Volcanic bomb   A blob of lava exploded from a volcano and spun through the air, then cooled and hardened. Bombs are larger in size than very fine ash or lapilli (see also *Volcanic ash* and *Lapilli*).

Volcano   A vent in the Earth through which molten rock issues or from which molten rock or volcanic gases are blown.

Wadi   See *Arroyo*.

Water cycle   See *Hydrologic cycle*.

Wind gap   A mountain pass not occupied by a stream. In many places, the stream no longer flows in the gap because another stream has captured its waters (see also *Stream piracy*).

Suggestions for Further Reading

BOOKS COVERING THE ENTIRE STATE

California's Changing Landscapes, by Gordon B. Oakeshott. 2d edition. Published by McGraw-Hill Book Company, New York. 379 pages, 1978. Available in paperback.

Geology of California, by Robert M. Norris and Robert W. Webb. Published by John Wiley & Sons, Inc., New York. 365 pages plus index, 1976. More technical than *California's Changing Landscapes*.

Evolution of the California Landscape, by Norman E. A. Hinds. Published as Bulletin 158 by the California Division of Mines (now the California Division of Mines and Geology), San Francisco. 240 pages, 1952. Well illustrated, but out of print and out-of-date.

BOOKS COVERING PARTS OF THE STATE

Geology of the Sierra Nevada, by Mary Hill. Published as California Natural History Guide 37 by University of California Press, Berkeley. 232 pages, 1975. Available in paperback.

The Incomparable Valley, a Geologic Interpretation of the Yosemite, by François E. Matthes, edited by Fritiof Fryxell, with photographs by Ansel Adams. Published by University of California Press, Berkeley, 1950. Available in paperback.

Sequoia National Park, a Geological Album, by François Matthes. Edited by Fritiof Fryxell. Published by University of California Press, Berkeley. 136 pages, 1950.

Death Valley: Geology, Ecology, Archaeology, by Charles B. Hunt. Published by University of California Press, Berkeley. 234 pages, 1975. Available in paperback.

Geologic History of Middle California, by Arthur D. Howard. Published as California Natural History Guide 43 by University of California Press, Berkeley. 113 pages, 1979. Available in paperback.

Earthquake Country, by Robert Iacopi. Revised edition. Published by Lane Books, Menlo Park, California. 160 pages, 1971. Available in paperback.

Underwater California, by Wheeler J. North. Published by University of California Press, Berkeley, as California Natural History Guide 39. 276 pages, 1976. Relatively little about landforms. Available in paperback.

GUIDEBOOKS

Geology Field Guide to Southern California, by Robert P. Sharp. Published by Wm. C. Brown Company Publishers, Dubuque, Iowa. 181 pages, 1972. Interestingly written guide, with general material on the natural provinces of southern California, as well as logs for field trips to Death Valley and to the Mammoth area. Available in paperback.

Geology Field Guide to Northern California, by John W. Harbaugh. Published by Wm. C. Brown Company Publishers, Dubuque, Iowa. 123 pages, 1974. Consists chiefly of field trips to four areas: Point Reyes, Santa Cruz Mountains, Mounts Lassen and Shasta, and Sierra Nevada and Mono Craters. Available in paperback.

Roadside Geology of Northern California, by David D. Alt and Donald W. Hyndman. Published by Mountain Press Publishing Co., 279 West Front Street, Missoula, Montana. 244 pages, 1975. A book to carry while traveling. Available in paperback.

Geologic Guidebook along Highway 49—Sierran Gold Belt. The Mother Lode Country, preparation directed by Olaf P. Jenkins. Published as Bulletin 141 by California Division of Mines (now California Division of Mines and Geology), San Francisco. 164 pages, 1948. Out of print and somewhat out-of-date but still good.

Geologic Guidebook of the San Francisco Bay Counties: History, Landscape, Geology, Fossils, Minerals, Industry, and Routes to Travel, preparation directed by Olaf P. Jenkins. Published as Bulletin 154 by California Division of Mines (now California Division of Mines and Geology), San Francisco. 392 pages, 1951. Out of print and out-of-date but still useful.

Geology of Northern California, edited by Edgar H. Bailey. Published as Bulletin 190 by California Division of Mines and Geology, San Francisco. 507 pages, 1966. More technical than the preceding guidebooks.

Geology of Southern California, edited by Richard H. Jahns. Published as Bulletin 170 by California Division of Mines (now California Division of Mines and Geology), San Francisco. 878 pages, plus many maps, 1966. Technical and long out of print but still useful.

Aeolian Features of Southern California: a Comparative Planetary Geology Guidebook, edited by Ronald Greeley, Michael B. Womer, Ronald P. Papson, and Paul D. Spudis. Published by Arizona State University, College of the Desert, and the National Aeronautics and Space Administration—Ames Research Center. 264 pages, 1978. A well-illustrated, up-to-date book on California deserts.

Index

Tor, 149, 188, 216–17, 238; photo of, cover; drawing of, 188
Torrey Pines Beach, 105
Transverse Ranges, 29–32, 142, 144
Travertine Point, 216, 218
Tree-ring chronology, 238
Tree-ring method of determining time, 9–10
Trinidad Beach, 178
Trinity Alps, 27, 107, 127; photo of, center
Trona Pinnacles, 217
Truckee, 110
Tsunami, 157, 238
Tuchek, Ernest T., ix
Tucson, Arizona, 206
Tufa, 191
Tufa cone, 107, 191–193
Tufa dome, 107, 191–193, 217, 238; photo of, center
Tuff, 55–56, 70, 73, 238
Tujunga, 93
Tulainyo Lake, 101
Tulare Lake, 100
Tule Lake, 100
Tule Wash, 110
Tuolumne Glacier, 113
Tuolumne Meadows, 130, 132
Tuolumne Table Mountain, 61–63, 74, drawings of, 62–63
Turtle Mountains, 74
Twain, Mark, ix, 197
Twin Falls, 107
Twin Lakes, 101, 123, 127, 130
Twin Peaks, 227, 228
Two Teats, 71, 132

U-shaped valley, 118, 134
U.S. Cavalry, 58
U.S. Geological Survey, vii, viii, ix, 56, 65, 123; drawing from, 12, 170, 173, 174
U.S. National Park Service, 65
Ubehebe Craters, 53, 71
Underwater California, 242
Underwater pinnacle, 179
Underwater reef, 179

Uniformitarianism, principle of, 3
Upper Yosemite Falls, 131
Urbanization, 222–230
Uruq dune, 208

ValVerde Park, 106
Valerie Jean, 108
Valley Springs Peak, 73
Valley glacier, 235
Valyermo, 153
Van Norman Dam, 151
Varnish, desert, see Desert varnish
Varve, 11, 238
Vasquez Rock, 105
Ventifact, 203, 220, 239; drawing of, 204
Ventura, 106, 178
Ventura Basin, 154
Ventura County, springs in, 105
Ventura River valley, 106
Verdugo Hills Cemetery, 93
Vernal Falls, 118
Vernal pool, 97, 239
Vesuvius volcano, 48
Victorville, 219, 220
Volcanic ash, 49, 55, 69, 73, 236, 239
Volcanic block, 49
Volcanic bomb, 35–36, 49, 69, 231, 239
Volcanic country, map of, 50
Volcanic dome, 51, 60, 70–71
Volcanic glass, 15, 51
Volcanic lake, 98, 102
Volcanic landforms, photo of, center
Volcanic mudflow, 43, 48–49, 71
Volcanic neck, 74
Volcanic rocks, 15, 48
Volcanic tablelands, 74
Volcanoes, 35–74, 239

Wadi, 197
Walker Lake, 101, 123, 130
Walking stones, 195–96, 221; drawing of, 196
Wallace, R. E., viii